Insects

Photographic Guide for Identification

Books by Same Author

- Agricultural Insect Pests and their Control 2nd Ed
  ISBN: 9789386237125
- Insect Neuroendocrines
  ISBN: 9788172333904
- Introduction to General and Applied Entomology
  3rd Ed
  ISBN: 9788172335977
- Principles of Insect Behaviour 2nd Ed
  ISBN: 9788172338251
- Shabd Devta (Hindi)
  ISBN: 9788172339258

Insects

Photographic Guide for Identification

Dr. V.B. Awasthi

SCIENTIFIC PUBLISHERS

Published by

SCIENTIFIC PUBLISHERS (INDIA)
5-A, New Pali Road,
P.O. Box 91
Jodhpur - 342 001 INDIA
E-mail: info@scientificpub.com
Website: scientificpub.com

Print 2025

ISBN: 978-93-88172-78-3
eISBN: 978-93-88172-80-6

Printed in India

Dedication

This book is dedicated to
Dr. H.S. Zim, E. Cotton
and
Dr. Goodden Whose
work forms the base of this book.

PREFACE

This book entitled "Insect at a Glance" includes the familiar insects of Asia, America, Africa and Australia. This insect collection was made and photographed by me in 1960. I would like to mention that since my schooling days, I was very much interested in collecting and studying insects due to their beauty and misterious ways of Life. I found that they appear when temperature and humidity are ambient and after a few months they disappear. Later, I found that in adverse Indian conditions undergoes hibernation or aestivation or diapause in larval, pupal or adult stages. In beauty and variety they are next to flowers.

Long ago I read "Insect A guide to familiar American insects" by Zim and Cotton (1956). I found that guide very useful when I was a student of entomology for correct and quick identification of insects, without consulting advance books due to their coloured and clear illustrations. Thereafter more than 62 years elapsed but no book of such a character was published for beginners and graduate students of entomology. I thought to compile the common insects in colour for beginners as well as graduate students of entomology of a large part of world.

I would like to state that except a few, a large number of same species of insects viz. butterflies, moths, bugs, beetles, blue and green bottle flies, fruit flies, horse flies, flower flies, dragon flies, ant lions, cockroaches, stored grain insects aphids, milk weed, bugs, stink bugs, honey bees, crickets, termites ants, mosquitoes, stick insects etc. are either cosmopolitan or common in Asia, America, Africa, Australia and several other countries with slightly different colour forms due to change in climatic conditions (winter forms/ summer forms). Hence, they have not been classified/categorized country wise in this guide. Infact, insects have no boundaries like man and are free to migrate from one country to another.

This book deals only with common important and showy insects of a large part of world. I hope the book will be useful for quick and correct identification of insects.

I also hope that the students not only study insects but also help in the conservation of these tiny beautiful creatures (only 1% are harmful). Which are facing great difficulty due to lack of water bodies, destruction of food plants and deforestation, global warming lack of proper ecophysiological living conditions, human interference as well as unwanted use of synthetic pesticides. Many species which were commonly seen a few decade ago now have totally disappeared from the globe. I am extremely thankful to my wife Sneh Lata for her constant encouragement and whole hearted support, I am particularly grateful to my daughter Shubha for reading the proof and suggesting necessary corrections.

Dr. V.B. Awasthi

ACKNOWLEDGEMENTS

I would like to mention very honestly that I was very much impressed by Zim and Cotton (1956) when I was young and number of illustrations that are still in good condition have been reproduced in this book for which I whole heartedly acknowledge. In addition Dr. Lefroy, Fletcher, Goodden, Blaney, Trivedy, Pradhan etc. whose books have a great source of illustrations and information which I gratefully acknowledge.

Handbook of Insects: Photographic Guide for Identification

By dealing only with common, important, and showy insects, this book will help the beginners as well as graduate students a fascinating study. To identify an insect, turn to the key to the insect groups (orders) on pages 2 to 4. This key may place your specimen within one of 18 groups. Thumb through the section indicated to find your specimen or something similar. Study all the illustrations to become aware of insects you might not otherwise notice. You may thus become able to identify some at first sight. Don't be disappointed with failures. It may take an expert a year or more, after finding on insect, to make a final identification. If you can place in their proper orders a quarter of the insects you find, you have made a good start.

Insects in this book are often shown on their food plants. Immature forms often appear with the adult. If you cannot identify an immature insect, try to raise it to maturity.

In advanced study the Latin scientific names of species are used for greater precision in designation. Scientific names of species illustrated in this book are given at the end of book.

On plates, approximate lengths are given in inches (``w". indicates width of wings).

KEY TO INSECT ORDERS

Grasshoppers Locusts and crickets (Orthopter) medium to large, mouth parts chewing type forewing narrow hind legs membrous broad, many veined, hingleys modified for jamping. Mostly plant feeds development gradual.

Cockroaches and mantids (Dictyoptera) Body depressed, head partly under prothorax, antennae long, filiform dark brown or black in colour foul smelling, nocturn, omnivorous, spoil food etc. In praying mantis, head free prothorax long fore legs grasping type Development gradual.

Leaf insects and Stick insects (Phasmida) **:** Stick or leaf like brown or grey or green colour mimiking twigs or leaves. Mesothorax metathorax and legs long. Ovipositor small, cereal short leaf eaters Development gradual.

EARWIGS (Dermaptera), Small insects with typical pinecer like tail. Usually four small wings. Segmented antennae. Development gradual.

TERMITES (Isoptera). Ant-like insects, small and soft bodied. Some have four long wings. Live in colonies. Specialized ``castes" for working, fighting. Chewing mouthparts. Development gradual.

LICE (Anoplura). Small, wingless insects with piercing and sucking mouth-parts. Body flattened. Legs with claws for clinging to warm-blooded animals.

LEAFHOPPERS, APHIDS, AND SCALE INSECTS (Homoptera). Small to medium insects, most with two pairs of similar wings held sloping at sides of body. Jointed beak for sucking attached to base of head. Land insects, Some scale-like.

TRUE BUGS (Hemiptera) Range from small to large in size. Two pairs of wings, with forewings partly thickened. Jointed beak for sucking arises from front of head. Development is gradual.

DRAGONFLIES AND THEIR KIN (Odonata). Fairly large insects with two pairs of long, equal sized wings. Body long and slender. Antennae short, Immature insects are aquatic. Development in three stages.

MAYFLIES (Ephemeroptera) AND STONEFLIES (Plecoptera). Both with two pairs of transparent, veined wings. In mayflies, hind wings are smaller, in stoneflies they are larger. Mayflies have long, 2 or 3-pronged tails.

NERVE WINGED INSECTS (Neuroptera). The two pairs of wings, usually equal in size, are netted with veins. Four stages of development, egg, larva, pupa, and adult. Chewing mouth parts. Long antennae.

SCORPIONFLIES (Mecoptera). Small insects with two pairs of slender, generally spotted wings. Legs long. Antennae long also. Beak like chewing mouth parts. Larvae live in soil.

CADDISFLIES (Trichoptera). Most larvae live in fresh water. Some build ornamented case. Adults with two pairs of wings with long, silky hairs and with long antennae. Mouth-parts reduced.

MOTHS AND BUTTERFLIES (Lepidoptera). Medium to large insects with two pairs of scaly wings. Sucking mouth parts. Antennae knob-like or feathery. Development in four stages.

FLIES AND THEIR KIN (Diptera). Two winged, small to medium insects, with sucking mouth parts. Antennae small, eyes large. Second pair of wings reduced to balancing organs. Development in four stages.

BEETLES (Coleoptera). Fore-wings modified to thickened covers. Hind wings thin, folded. Size from small to large. Chewing mouth parts. Antennae usually short. All have four life stages. Some aquatic.

BEES, WASPS, AND ANTS (Hymenoptera). Small to medium size insects many social or colonial. Two pairs of thin, transparent wings. Hindwings smaller. Mouthparts for chewing or sucking. Only insects with ``stingers''. Development in four stages.

SEEING INSECTS

One cannot avoid seeing insects, as they are found everywhere, even in the Antarctic. They have been on this earth some 250-500 million years ago, and seem here to stay. More insects and more kinds of insects are known than all other animals visible to the naked eye. Insects have been called man's worst enemey. Certainly, they are, but some have economic value, and for other reasons we would be hard put to exist without them. Insects are gems of natural beauty, zoological mysteries, and a constant source of interest.

WHAT INSECTS ARE

Insects are related to crabs and lobsters. Like these sea animals they possess a kind of skeleton on the outside of their bodies. The body itself is composed of three divisions; head, thorax, and abdomen. The thorax has three segments, each with a pair of jointed legs; so an insect normally has six legs. Most insects also have two pairs of wings attached to the thorax, but some have only one pair, and a few have none at all. Insects usually have two sets of jaws, two kinds of eyes simple and compound - and one pair of antennae.

So much for the typical insects, but many common ones are not

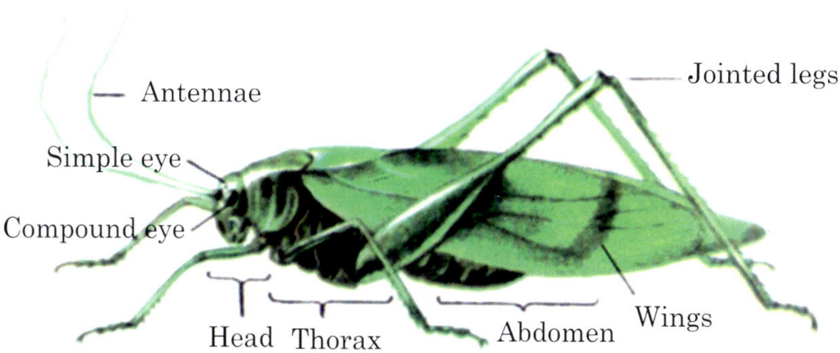

PARTS OF AN INSECT

typical. The thorax and abdomen may appear to run together. Immature stages (larvae) of many insects are worm-like, though their six true legs and perhaps some extra false ones may be counted. Immature insects are often difficult to identify. It is also hard to tell the sex of some insects but in most cases, sexual dimorphism is distinct. In some groups males are larger or have larger antennae or different markings.. The female is sometimes marked by a spear-like ovipositor for laying eggs extending from the base of the abdomen.

INSECT RELATIVES

A number of insect-like animals are confused with insects. Spiders have only two body divisions and four pairs of legs. They have no antennae. Some spiders do not live in webs. Other insect like animals have the head and thorax joined like the spiders. Crustaceous have at least five pairs of legs and two pairs of antennae. Most live in water (crab, lobster, shrimp), but the lowly sowbug is a land crustacean. Centipedes and millipedes have many segments to their bodies with one pair of legs (centipedes) or two pairs (millipedes) on each. Centipedes have a pair of long antennae; millipedes have a short pair. Millipedes often coil up when disturbed.

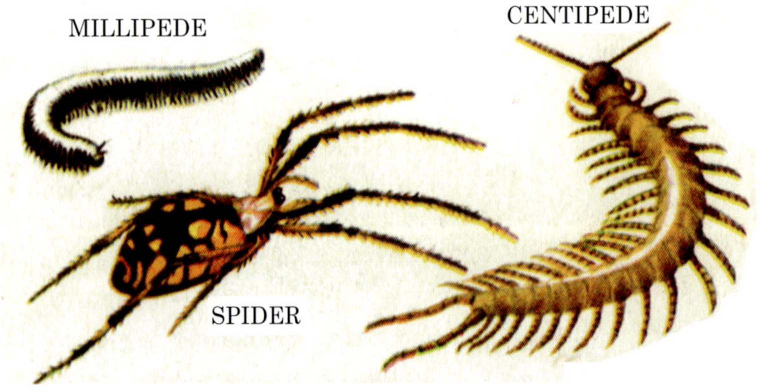

NUMBER OF INSECTS

The insect group is by far the largest group of animals in the world. It has recently been estimated that the total number of insect species may range from 4-10 million (See : Atwal and Dhaliwal, 2012). The class of insects are divided into some 28 orders (depending on the classification). Five ``major'' orders stand out for their high species richness, the beetles (Coleoptera), flies (Diptera), wasps, ants, and bees (Hymenoptera), butterflies and moths (Lepidoptera), and the true bugs (Hemiptera). Beetles - appears to be confirmed since they comprise almost 40% of described insects (more than 350,000 species). The Hymenoptera have nearly 250,000 described species, with the Diptera and Lepidoptera having between 125,000 and 150,000 species, and Hemiptera approaching 95,000. Of the remaining orders of living insects, none exceed the 20,000 described species of the Orthoptera (grasshoppers, locusts, crickets and katydids). Dermaptera less than 2,000 described species and cockroaches (Blattaria) with only 4,000 species. Most of the ``minor'' orders have from some hundreds to a few thousands of described species (Gullan and Cranston, 2005).

INSECTS AND MAN

Whether certain insects are to be considered helpful to man or otherwise, depends as much on man as it does on insects. Our ways of farming and raising animals have provided some insects which might otherwise be rare with conditions enabling them to multiply a thousandfold. Only about 1 percent of insects are harmful, but these destroy about 10 percent of our crops. Other insects are parasitic on other animals, and some are known to carry diseases.

On the other hand, this would be a sorry world without insects. We would have no apples, grapes, or pear much less cotton, and fewer oranges and garden vegetables, for these and many other common plants depend on insects to pollinate their flowers. And there would be no honey, of course. Some insects aid the process of decay, a process that is essential to life. Some insects help control others, and all help maintain a balance in nature.

INSECTS IN THEIR PLACE

In the broad view, insects play an important natural role, not only in ways that benefit man but as food for many kinds of fish, amphibians, birds, and mammals. Many of our songbirds depend almost entirely on an insect diet. Insect help make our rich plant life and wildlife possible.

CONTROL OF INSECTS

There are ways to supplement the natural control of insects by birds and other animals. The modern synthetic insecticides are both effective and reliable. Most of the phytophagous and stored grain insects can be controlled by use of suitable synthetic insecticides.

Systemic insecticides viz phorate, dimethoate, phosphomidon, quinolphos etc. are also used against sap feeding and leaf eating insects. These are used as dust, granules, wettable powders, solutions and emulsifiable concentrates. These insecticides must be used under the supervision of experts because all these are extremely poisonous. Attractants, repellants, Atlifeedants, fumigants are also used for the control of various insects.

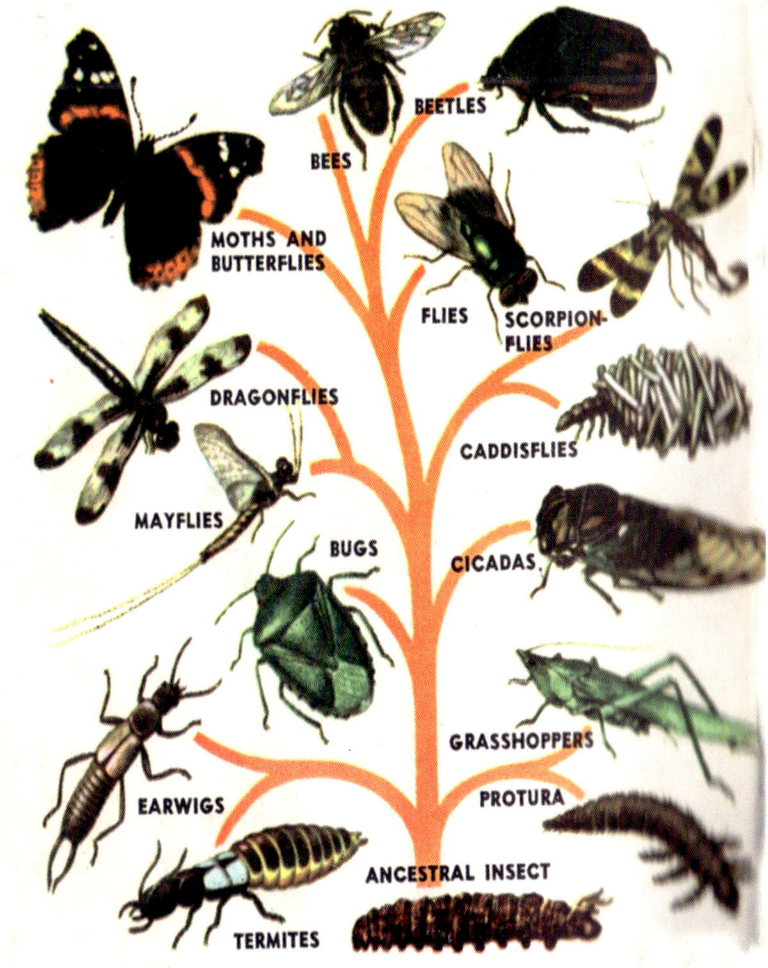

FAMILY TREE OF INSECTS (From Zim and Cottam, 1956)

FAMILY TREE OF INSECTS The ancestor of all insects was probably a segmented warm-like creature much like primitive protura, silverfish, springtails, and kin. As long as 500 million years ago. Cockroaches and other insects were common. Today there are 28 or more orders of insects (depending on the classification). Most of the 12,000 kinds of fossil insects identified are similar to living species. For relationships of insects, see ``family tree'' above.

SILVERFISH – primitive insect with direct, gradual development

CHINCH BUG – more advanced insect, with incomplete metamorphosis or development in three stages

HOUSE FLY – advanced, modern insect, with complete metamorhosis or development in four growth stages

Insects follow different developmental patterns. In the simplest, the newly hatched insect is like a miniature adult. It grows and moults (sheds its skin) till it reaches adult size. In *incomplete metamorphosis* an immature nymph hatches, grows, develops wings, and by stages becomes an adult. *Complete metamorphosis* involves (1) egg, (2) larva, (3) pupa or resting stage, (4) adult.

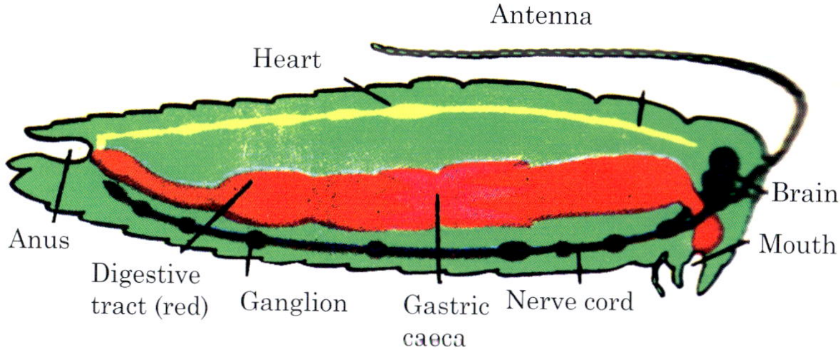

INTERIOR OF A GRASSHOPPER (Side view)

INSECT STRUCTURE is marked by three body divisions, six jointed legs, one pair of antennae, and usually one or two pairs of wings. The outer covering, or exoskeleton, is often horny. Mouth parts are complicated. Internally, insects possess many of the organs. They have a digestive tract and auxiliary digestive organs. Breathing is done by air tubes spreading internally from openings called spiracles. The heart is tube like; blood circulation is simple. Respiration, digestion, and circulation are shown in the longitudinal section (above and cross section (left) of the grasshopper, a typical insect. The nervous system below) shows the brain, which exerts control over the body. Ganglions serve as nerve centers for nearby parts of the body. Excretory, reproductive and endocrine organs have not been shown in this figure.

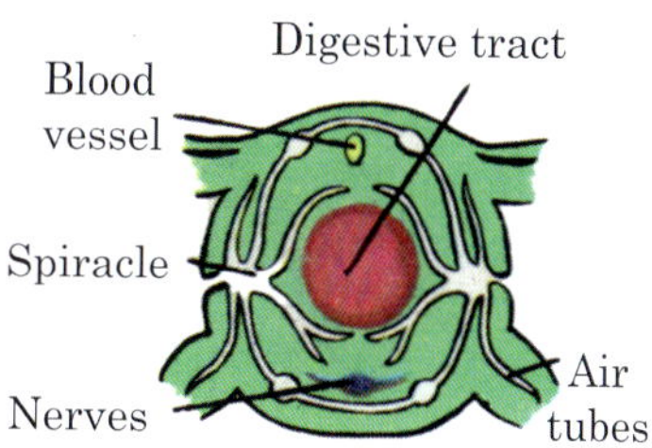

CROSS-SECTION

NERVOUS SYSTEM
showing brain, nerves, and ganglions

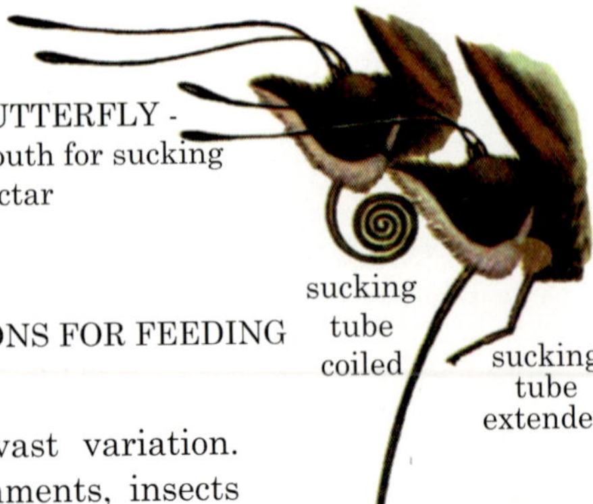

ADAPTATIONS FOR FEEDING

Insect structures show vast variation. Adapted to many environments, insects live successfully in nearly every part of the world. They have digestive systems for all kinds of plant and animal food. They thrive on everything from wood to blood. A few species do not eat at all in the adult stage. Mouth parts are adapted for chewing, sucking, piercing and sucking, and lapping. Equally interesting adaptations are seen in insect wings, body coverings, and reproductive organs. The typical insect leg (e.g. grasshopper) has five parts. Grasshopper hind legs are specialized for jumping. The house fly has pads which enable it to walk up windows. In honeybees, the hind leg is adapted to store and carry pollen, and the foreleg to groom and clean antennae and body. Insect structures make a fascinating study.

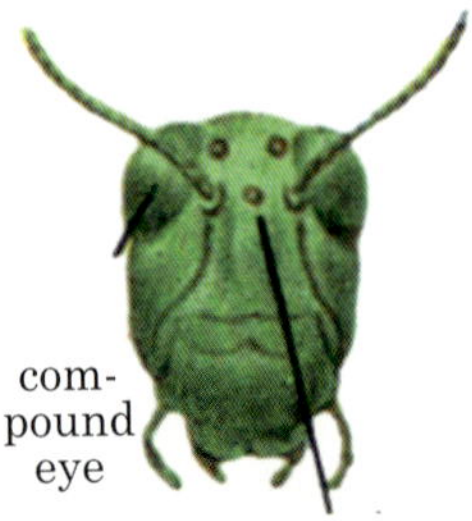

ADAPTATIONS OF FEET

STUDYING AND COLLECTING INSECTS

WHERE TO LOOK Practically everywhere : in fields, gardens, woods, roadsides, and swamps; under stones, rotted logs and leaves. Look in flowers, on grass; on animals, too. You'll find insects in the air, in and on water; an and in the ground.

WHEN TO LOOK Insects are most common in spring and early autumn but experienced collectors can find them in all seasons. Some groups of insects are more common at night remember that when collecting. In winter, concentrate on protected spots, as under stones and bark. Watch for insects in egg cases or in the resting stage (pupae).

WHAT TO DO Studying insects is not confined to catching them and mounting them in collections. Raising live insects to study their habits is exciting.

Watch carefully how the caterpillars shed the skins, spin a cocoon or chrysalis, and finally emerge as a moth or a butterfly. See warm-like larvae become flies or beetles. You will learn more from live insects than from dead specimens.

Whatever you do with insects, you will need some understanding of what insects are and how they live. Use this book, then read some of the other advance books. Most important, go out and look at insects. Catch them if you wish, but watch them first. See how they move, how they feed and what they do.

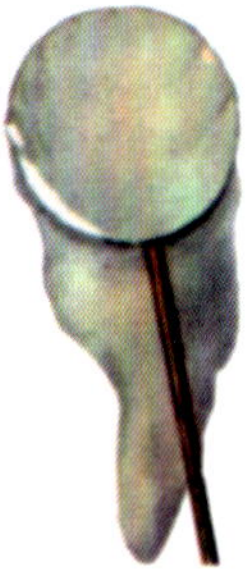

COLLECTING INSECTS An insect collection can be valuable for study or reference if it is used. Beginners can collect insects with simple, low cost equipment. Not all insects are chased with a net. Hang an old bedsheet out at night with a small light in front of it. Roll the bottom into a funnel, with a jar set beneath. Insects hitting the sheet will roll down into the jar. Similar traps are described in advance books.

THESE INSECTS WERE COLLECTED BY THE AUTHOR IN 1962

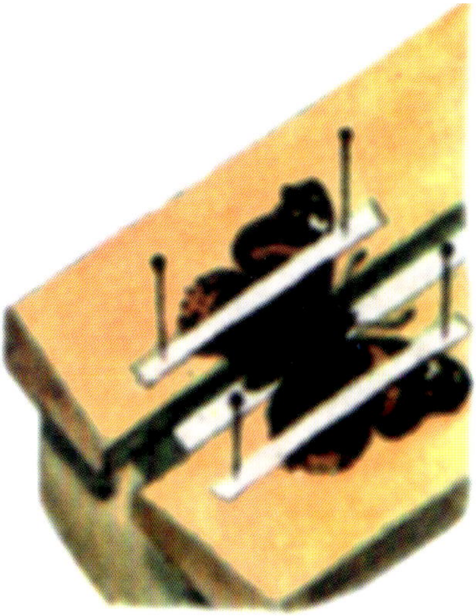

EQUIPMENTS Any large, wide mouthed jar will serve for confining and raising insects. Tie some gauze or netting over the top. To kill specimens use a wide mouthed bottle with absorbent cotton or sawdust on the bottom. wet with a tablespoon or so of carbon tetrachloride or carbana. Cover this with a sheet of tight-fitting cardboard punched full of pinholes. Because cyanide is such a dangerous poison, cyanide bottles should be made and used only by experienced collectors. A light net with a long handle is good for catching insects on the wing. A heavier net is better for ``sweeping'' through the grass or for catching water insects.

Plastic or wooden glass boxes with a layer of heavy corrugated cardboard on the bottom to hold pins are fine for storing specimens. Use a mothball to deter insect pests. Use insect pines; ordinary ones are too heavy. Learn the tricks that make mounting neat and attractive. A note book for records is essential; so are labels. Later you may want spreading boards, pinning blocks, and other accessories. Collecting and preserving specimens requires real skill. Read first then practice with any insects you may find in your own yard. Skill will come with experiences.

LIFE HISTORY STUDIES This may prove more interesting and exciting than collecting. Instead of leaning a little about many insects, learn a lot about a few. Many insects are known only in the adult form; few facts are known about the rest of their life cycles. Constant observation of wild specimens or detailed study of captive ones reared under natural conditions, may yield many new and interesting facts about insects.

LEAF INSECTS 2-3"

LEAF INSECT The leaf insects are fairly large in size. The wings of leaf insects are exactly leaf like. They feed on leaves and remain un-noticed when present on long leaved plants. They are camouflaged well due to their colour and are found in South America, Africa, Malaysia and Australia.

WALKINGSTICKS The stick insects are large, usually, wingless insects with legs all about the same length. Walkingsticks live and feed on leaves of oak, cherry, and walnut, occasionally causing damage. The female's 100 or so eggs are dropped singly on the ground to hatch the following spring. As young grow, they moult or shed their skin five or six times; they are similar to adults. Males are smaller than females.

WALKINGSTICK 3.0"

KATYDIDS The male of the true katydid makes the persistent ``katydid" call, which, coming as often as once a second, is an accepted part of a summer's evening. Some katydids are free dwellers, feeding on leaves of cherry and apple. Others live in the grass. Most are green, with thin, leaf like wing covers, and so have the advantage of protective coloration. All have long antennae. At the base of the outer wings or wing covers of the males are rasps and ridges which, when rubbed like a fiddle and bow, produce the calls of the different species. Katydids hear by ``ears" on the upper part of their front legs.

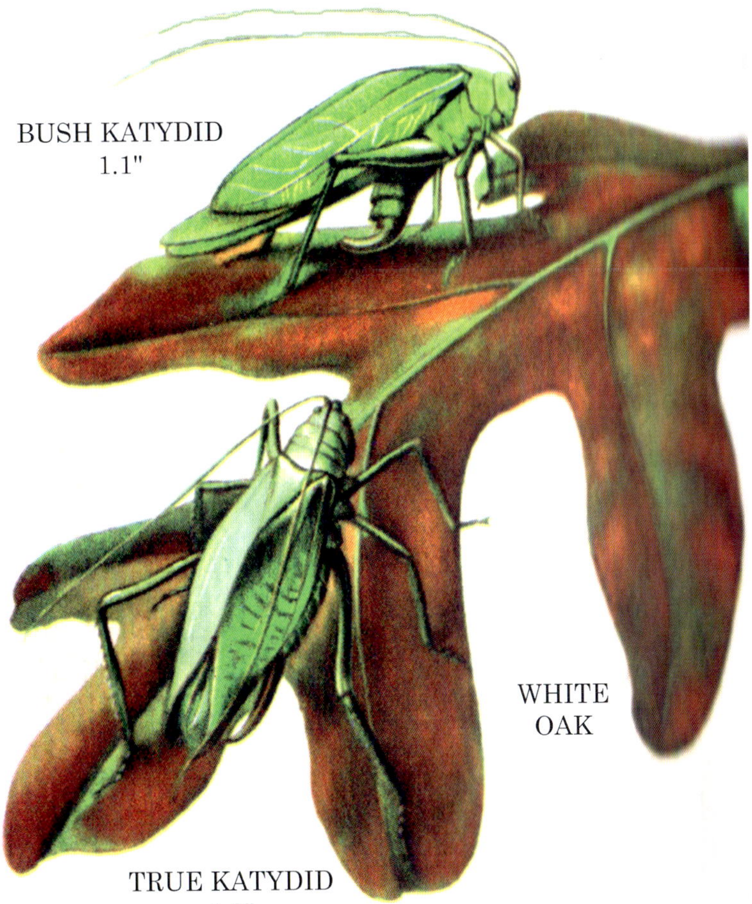

The females, recognized by the long ovipositor, lay their oval eggs on leaves or twigs early in the fall. About 100 to 150 are laid. When the young emerge in the spring, they resemble their parents, but are much smaller, lighter in colour, and lack wings.

MOLE CRICKET 1.3"

CAMEL CRICKET 0.8"

MOLE AND CAMEL CRICKETS These nocturnal crickets live under rocks in moist places or mostly underground. The large mole cricket burrows near the surface, eating young roots and killing seedlings. In the South America, it destroys peanuts, strawberries, and other garden crops. The pale brown, spotted, wingless camel (or cove) cricket is identified by its high, arched back. Though a scavenger, it often becomes a nuisance around green house.

AFRICAN MOLE CRICKET 1.5-2"

AFRICAN MOLE CRICKET This mole cricket is cosmopolitan in tropical part of Asia, Africa, Mauritius, Australia and Pacific islands. General characters are similar to other mole crickets.

MORMON CRICKET female 1.5" male smaller

MORMON CRICKET This serious pest of Western grains and other crops is partly controlled by insect parasites, small mammals, and birds. The large clumsy, insects devour everything in three path, including each other. Some Western Indians considered them a delicacy and ate-them roasted. Small clusters of eggs are laid in the ground by the female.

FIELD CRICKETS These common, large headed black or brown crickets are nocturnal. Their shrill musical night song is made by rubbing the forewings. Though vegetarians, they may eat other insects and each other. In captivity, these crickets make fine pets. Field crickets damage crops and occasionally invade homes, even eating clothing. Eggs are laid in the ground in fall. The young nymphs emerge in the spring and develop their adult wings in several stages.

FIELD CRICKET 0.9"

TWO SPOTTED CRICKETS

These crickets are fairly large black in colour. They produce very loud and musical chirping. This is the most abundant cricket species which inhabits the tropical and subtropical regions of Asia, Africa and Europe. The other behavioural characters are similar to field crickets.

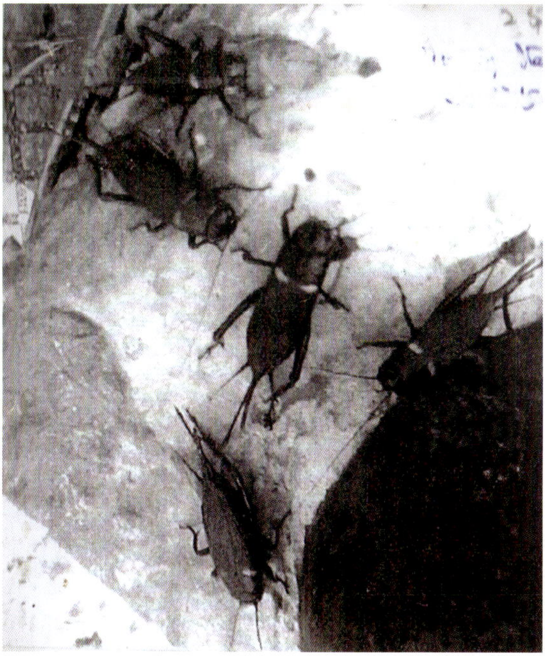

TWO SPOTTED CRICKETS 1"

PRAYING MANTIS

This mantis is very common in India and the neighbouring countries. It is found mostly on the peepal tree (*Ficus religiosa*) and other trees of the same family. All mantis are predacious and have identical characters.

TWO SPOTTED CRICKETS 3-4"

PRAYING MANTIS 3.5"

Lutz says: These are the only insects that can look over their shoulders. In autumn after mating, the female may eat the male. She lays several hundred eggs in a frothy mass that dries like hardened brown foam. Egg cases can be found in winter and brought indoors to hatch. The young similar to adults but light yellow, are difficult to raise.

GRASSHOPPERS AND LOCUSTS These are a group of closely related insects. Some species migrate; others do not. All are similar, with short antennae and large hearing organs on the abdomen. Most are good fliers, though some kinds are wingless. Locusts and grasshoppers destroy crops, but they serve as food for larger birds, small mammals, and other animals. Females lay 20 to 100 eggs in the ground or in rotted wood for the life history of one species. The nymphs mature in 2 to 3 months.

LUBER GRASSHOPPER
2.5"
MIGRATORY
GRASSHOPPER
1.3"
Wheat
Sun-flower leaf

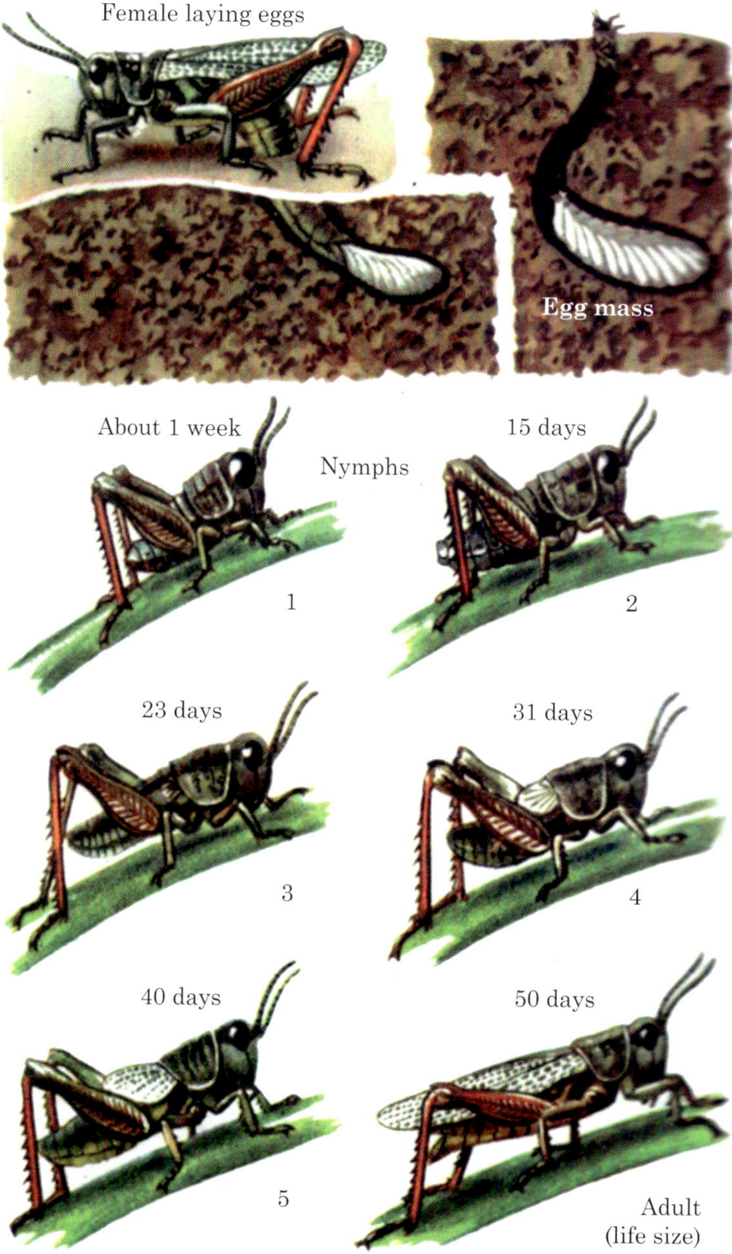

RED-LEGGED GRASSHOPPER

DESERT LOCUST 2-2.5"

DESERT LOCUST

Desert Locust is a polyphagous pest and virtually all crops are at risk so far as locusts are concerned but there is some preference of cereals. The leaves and soft shoots are eaten from margin inwards leaving irregularly shaped feeding marks. Swarm damage usually results in complete defoliation of crops. A swarm of desert locust may contain about 50 billion insects.

AK - GRASSHOPPER 2"

COCKROACHES

The most common household insect pests are the cockroaches. Cockroaches are generally brown or dark brown in colour. They are broad and flattened, with the head bent under the body so that the mouth opens backward and the eyes look downward. The antennae are long slender and of many segments. Wings are usually developed in adults. The mouth parts are strong. The legs long and in most species bear numerous spines. Cockroaches are active at night, hiding in dark places such as cracks and crevices during day light and running very rapidly when disturbed.

The cockroaches consume foods and food materials freely, gnaw woollen goods, leather, and anything that has paste on it, and thus often injure book bindings. In fact, they are practically omnivorous. Besides eating, they leave a disagreable odour which spoils food where they have been. They lay their eggs in pockets or capsules or ootheca, the number of capsules vary with species and the outside of ootheca is horny in nature. The nymphs are active, feed freely and moult several times. Several types of cockroaches are found in buildings, the most common are :

AMERICAN COCKROACH This is the largest of the house cockroaches. It is dark reddish brown with a more or less definite yellow band around the margin of the pronotum. Both males and females are winged bearing long active antennae. As many as 16 nymphs can hatch from an ootheca, incubation taking less than two months to complete. The nymphs are wingless but otherwise resemble the adults except for size. A generation requires ten to twenty months to develop which is also influenced by temperature. Adults are long lived and females may produce as many as 50 capsules in her life time.

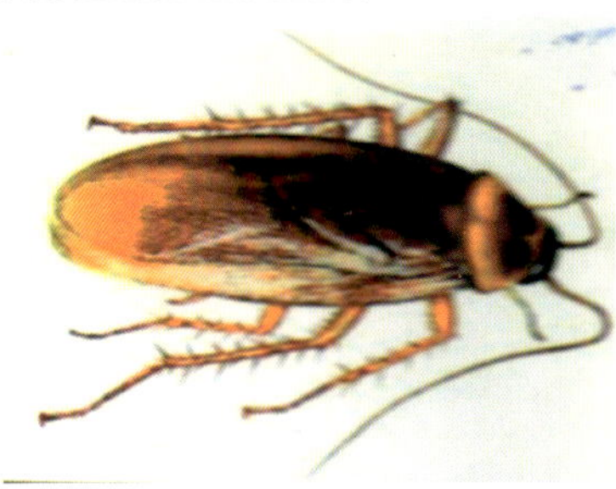

AMERICAN COCKROACH 1.5"

ORIENTAL COCKROACH

This is more commonly called 'black beetle'. It is almost black. The female is oval and has only wing stubs. The males have developed wings nearly covering the abdomen, are more slender.

The eggs capsules, containing a maximum 16 eggs are dropped as soon as formed and often are glued to objects. The time required for its young to develop is about the same as for the American Cockroach.

ORIENTAL COCKROACH 1.5"

BROWN-BANDED COCKROACH

These are small cockroaches bearing light tan band across the base of the wings and a broken band across the centre of the wings. The male has wings covering the tip of its slender abdomen, while the female is broad with stubby wings, egg capsules are small and tan in colour.

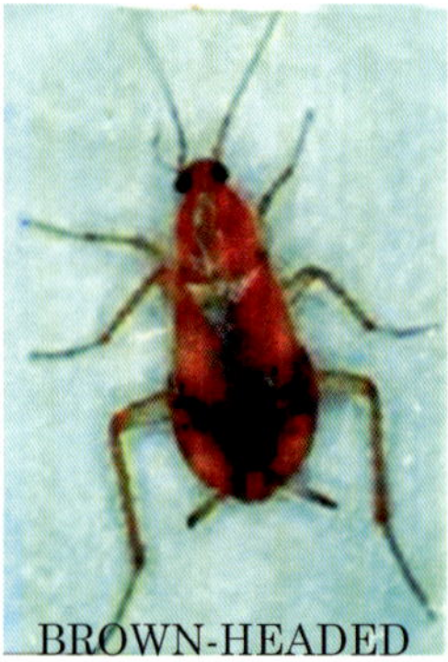

BROWN-HEADED COCKROACH 1"

The capsules of this species can have as many as 18 eggs each and hatching occurs in about 2 months. There are two generations in a year.

GERMAN COCKROACH It is pale brown in colour with two parallel, dark stripes on the pronotum. The adults are long, boat shaped in appearance and posses fully developed wings covering the abdomen. The antennae are longer than the body. The egg capsules or ootheca are slender and light brown in colour.

GERMAN COCKROACH 1"

EARWIGS

Euborella annulipes 0.8"

Labidura riparia 1"

Forticula auricularia 1.5"

EARWIG 0.6"

EARWIGS : These are marked by short, leathery forewings and a pincer-like abdominal appendage which is more pronounced in males. They are nocturnal spending the day in crevices or damp places. Some kinds are carnivorous, feeding on other insects. Young nymphs are wingless and gradually develop adult form.

TERMITES Though sometimes called white ants, termites are not ants, and some are not white. Of some 2,000 species, only about 40 are found in America and rest of them found in Asia, Australia and Africa. These highly socialized insects live in colonies composed of four distinct castes. The king and the queen, and the winged termites which can become kings and queens of

A COLONY OF TERMITE

new colonies, form the first caste. The enlarged and almost helpless queen produces thousands of eggs. Most of these hatch into whitish, blind workers who make up the second caste. Soldiers with large heads and jaws, and nymphs which take over the task of reproduction should the king or queen die, make up the last two castes.

The young pass through six stages in two years as they develop into adults. Tropical termites build huge nests or mounds, often higher than a man.

LICE are minute, wingless, disease-carrying insects that live and breed on their hosts. All are parasites. Biting lice (bird like), a distinct group, feed on hair, feathers, and fragments of skin. The sucking lice take the host's blood directly, by means of sucking mouth parts. The hag lice are the largest of this group. The head louse infects humans and is known to carry typhus, trench, and relapsing fever. Six to 12 generations of lice may mature annually. Young, similar to adults, develop rapidly.

HEAD LOUSE 0.1"

SHORT-NOSED CATTLE LOUSE 0.1"

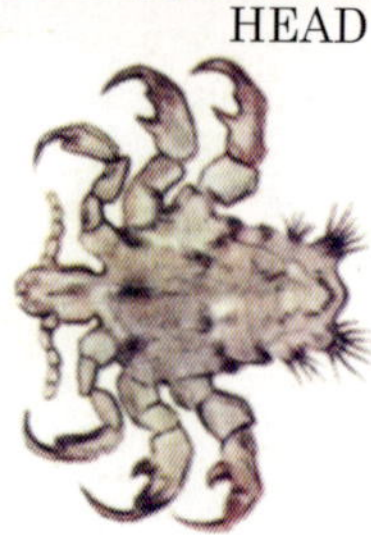

CRAB LOUSE 0.1"

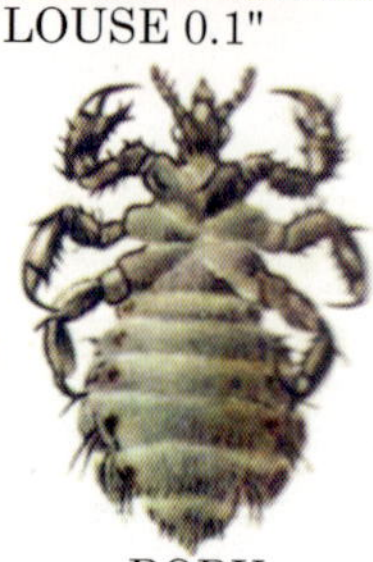

BODY LOUSE 0.1"

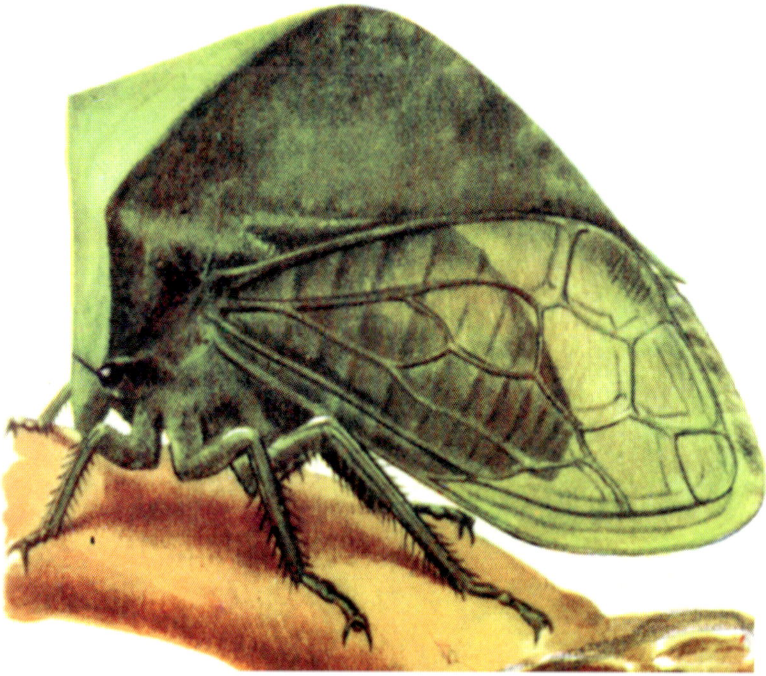

BUFFALO TREEHOPPER 0.4"

TREEHOPPERS The common green and brown tree hoppers are small, winged, sucking insects of curious and peculiar shapes. They live on many plants, feeding on the sap. Because of their protective colour and form, they are usually noticed only when moving. Eggs are laid in stems and buds, sometimes causing minor damage. Eggs hatch the following spring. Young are similar to adults.

LEAFHOPPERS These attractive, slender, multicoloured insects are often abundant on plants where they can feed by sucking the sap. This causes wilting and injury to grape, apple, beet, and other plants. Besides, leafhoppers carry virus diseases from plant to plant and thus become serious pests.

Leafhoppers exude ``honeydew'' as they feed. This is a somewhat sweet surplus sap which attracts ants and bees, which feed on it. Leafhoppers are well known as prodigious jumpers.

RED-BANDED LEAFHOPPER 0.3"

LATERAL LEAFHOPPER 0.4"

POTATO LEAFHOPPER 0.3"

3-BANDED LEAFHOPPER 0.3"

ROSE LEAFHOPPER 0.3"

The female lays eggs in stems and leaves. Two or more generations are produced each year. Late eggs winter over and hatch in spring. Adults hibernate and emerge in spring also. The young that hatch resemble the adults and pass through 4 or 5 nymph stages before they mature. Leafhopper populations in fields may reach as high as a million per acre.

CICADAS The cicadas, whose steady hum fills the late summer air, are more often heard than seen. Males make the sharp sound with plate like organs on the thorax. Some species are called harvest flies because of their late summer appearance; others are called 17 year locusts, though the 75 species of cicadas differ widely in the time they take to mature. The females cut slits in young twigs and deposit eggs in them. This habit causes damage in nurseries and orchards, because the slit twigs break easily in the wind. As the wingless, scaly young hatch, they drop on the ground, burrow in, and stay there 4 to 20 years (depending on the species and the latitude) as nymphs living on juices sucked from roots. The full grown nymph climbs a tree trunk. The skin splits down the back; the adult emerges. ``Broods" or large colonies may emerge *en masse*. Adults ordinarily live about a week long enough to mote and start another.

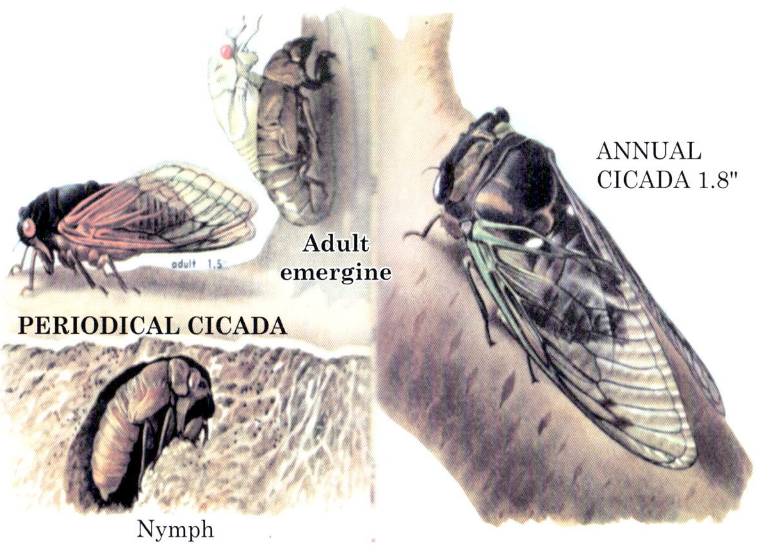

SPITTLEBUG The female spittlebugs make a froth on stems and grasses to cover their eggs. The young nymphs make a froth also to cover themselves while feeding. Spittlebugs are also called frog hoppers because the adults hop about from plant to plant and seldom fly. Though spittlebugs are of minor importance, some kinds injure pine trees and various garden plants.

APHIDS

The adult insects are very small having light green or slightly yellowish colour. They have a soft body and are of oblong in shape. One of the special morphological features in these insects is the presence of a pair of short tubes known as honey tubes which project from the upper surface of the hind end of the body. It is through these tubes that honey dew is excreted out. Generally apterous forms are present but during migration of the insect species winged forms are produced. They damage cabbage, radish, cauliflower etc.

APHID 0.2-0.5"

SAN JOSE SCALES

The pest is world wide in distribution. Several hundred species of other fruits as well as ornamental plants are severely damaged by this insect. It mostly prefers plants of the family Rosasae such as apple, plum, pear and other related species. The damage is caused by nymphs and the female scales which suck the sap form twigs, branches and fruits. The infested fruits also have scaly appearance ant each spot is surrounded scarlet or read area.

SAN JOSE SCALES 0.2-0.3"

LAC INSECTS 0.1"

LAC INSECTS

Lac is a resinous secretion of this insect. The insect feeds on certain tropical trees and is so injurious to its hosts that rather exacting methods must be followed in order that adequate crops of lac be obtained. The chief host trees are Kusum. (*Scheichera trijuga),* Palas *(Butea monosperma*). Ber (*Ziziphus jujuba),* Peepal (*Ficus religiosa*), Khair (*Acacia catechu*) etc.

Lac is used in preparing a large number of articles mainly by jwellers for filling hollow gold ornaments, bangles and bracelets etc. It is a very essential ingredients of paints and varnishes, and polishes. Lithographic ink, sealing wax. Shellac is used in gramophone records, buttons, pottery, artificial leathers etc.

SCALE INSECTS

Scale insects attack and injure citrus, apple, and other fruit trees, and greenhouse and ornamentals plants. Scales are difficult to control. Ladybugs, certain small wasps, and other natural enemies are used in fighting them.

Reproduction is complicated. Most scale insects spend the winter as eggs, which the female deposits under her shell before she dies. The eggs hatch in spring and the young move to fresh growth before they settle down under a scale. Some species produce several generations of females before normal sexual reproduction takes place.

STINKBUGS AND SHIELDBUGS There are several hundred species of stinkbugs and shieldbugs. All have the flattened, shield-shaped body. Most suck plant juices but some feed on other insects. Most are coloured green or brown, to match their environment, and are not easily noticed. A black species common on black berries and raspberries is so well concealed it is sometimes eaten. The colourful harlequin bug is an exception, even to its unusual eggs. The young, hatching from the eggs, pass through a series of growth stages till the nymphs become adults. The odour, which comes from two glands on the thorax and which gives stinkbugs their name, is also characteristic of a number of other bugs. Birds are not bothered by the odour and commonly feed on the insects. The harlequin bug and several other species are destructive to garden crops. The shieldbugs are very similar to stinkbugs. In these species, the shield, which develops from the thorax, is so large that it covers a good part of the abdomen.

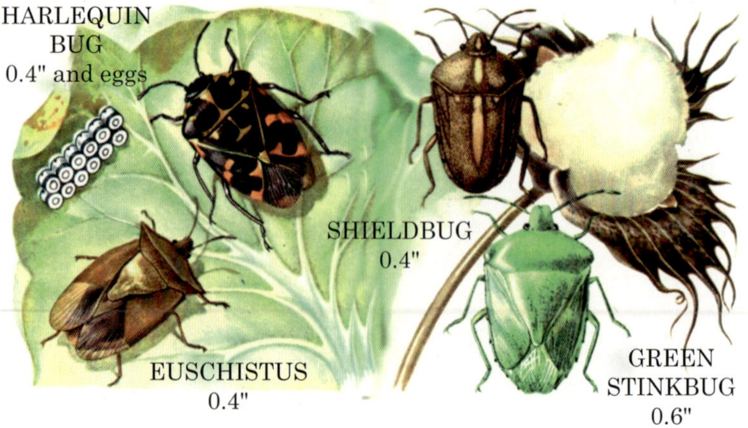

SQUASH BUGS They cause considerable damage to squash, pumpkins, gourds and related crops by sucking juices from leaves and stems of young plants. The bugs have a strong, offensive odour. Eggs, laid in late spring, hatch in about 2 weeks. The attractive nymph is green, soon turning brown or gray. Adults hibernate over the winter. The tachnid fly, which lays its eggs in nymphs and adult squashbugs, parasitizes these pests and helps in biological control of pests.

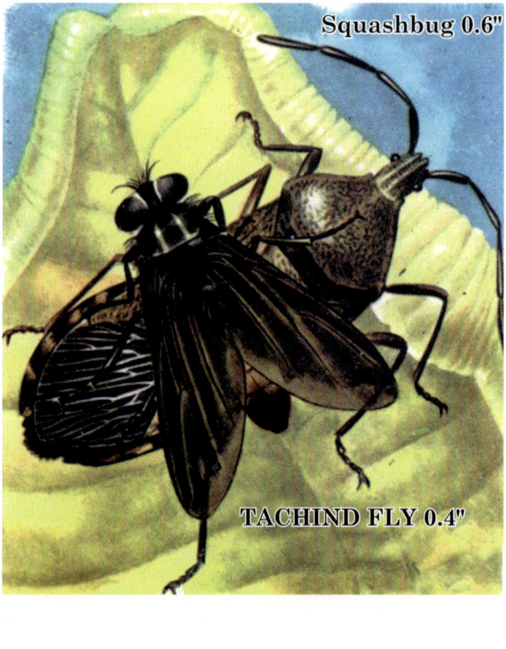

PAINTED BUG 0.5"

PAINTED BUG

It is a serious pest of many cruciferous plants as well as those of the cucurbitaceous species, the seriousness being sporadic. All the stages in the insects life cycle cause damage to the plants. The individuals suck the sap from the leaves which consequently turn yellow and drop off exposing the plant to bacterial and fungal infection. The adult is black in colour painted with irregular red markings. The scutellum is very large. The adults are extremely tenaceous and can withstand the absence of food for a fairly long period.

GUNDHI BUG 1"

GUNDHI BUG

It is a common pest of rice and is a major one. Both the nymphs and adults destroy the ripening grains of the plants usually during August to October, extracting their milky juice with the consequence that the ears turn white forming shriveled grains and preventing their maturation. A serious attack results in the appearance of empty, dry, brown ear heads on the plant. Yield losses of 10-40% are common and in severe infestations the entire crop may be destroyed. Plants such as those of jowar, maize, bajra, etc. form the secondary host of this insect.

GREEN POTATO BUG 0.6"

PUMPKIN STINK BUGS 0.6"

The adult insect is slender, green or light brown in colour. It emits strong unpleasant aromatic odour and hence the name `Gundhi bug'. Both male and female individuals are identical in appearance except that the abdominal apex of the latter is more swollen than that of the former.

RED COTTON BUG

The common name is ``red cotton bug''. It is a serious pest of cotton and sucks juice with its long beak from the seed of either green or ripe cotton bolls. Both the nymphs and adults are voracious feeders and cause extensive damage to the plant. The main symptoms of attack by this pest is the shrivelling of bolls in the plant. A severe infestation by this pest causes serious impairment to the vitality of the plant and destroys the germinating power of the seed by the loss of its oil contents. Besides these features, the formation of lint is badly affected.

RED COTTON BUG 0.6"

LITCHI BUG It is a pest of litchi, banana *Logan, Lantana* and a number of berries. The bug on emergence is red in colour. The female bug lays about 22 spherical chreamy white egg 1 or 3 batches. The nymphs emerge in 5-7 days. Which moult five times and completes life cycles in 25-40 days depending on climetic condition. The nymphs and adults suck the sap from young fruits causing withering of fruits.

LITCHI BUG 0.6-0.7"

MILKWEED BUGS These black and red or orange bugs are similar and closely related to the tiny destructive chinch bug. About 200 or more species are grouped in the same family with the milkweed bugs, but most are much smaller and less attractive in colour and pattern. Milkweed bugs feed on all varieties of milkweed and are of no economic importance. Adults which hibernate over the winter produce young in late spring. The nymphs mature and breed by late summer.

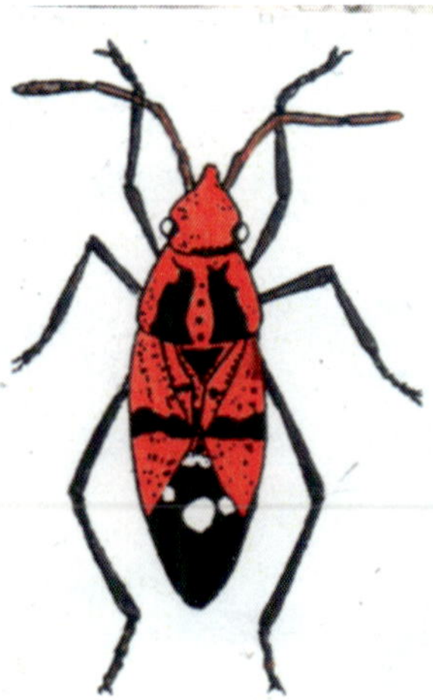

MILKWEED BUG 0.4-0.5"

Small Milkweed Bug 0.4"

Large Milkweed Bug 0.5"

Ambush Bug 0.4"
Feeding on aphid

AMBHUSH BUG These small, oddly shaped predators form a minor group. Their habit is to lie concealed in flowers, grasping any small insects which may come by. Ther front legs are modified for holding their victims; the mouth, for tearing and sucking. Interesting because of their bizarre forms and feeding habits, the ambush bugs lack economic importance.

CHINCH BUG Though small, almost minute, chinch bugs reproduce so rapidly they overrun grain fields, destroying the crops as they feed on plant juices. About 500 eggs are deposited in grass or grain. Nymphs are red, becoming gray or brown with age. Two or three generations may develop in one season. The furnished plant bug. somewhat larger, and of a related family, is destructive to many kinds of fruits.

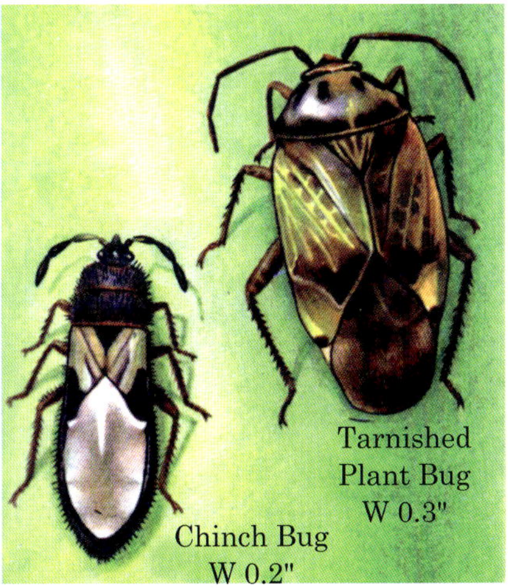

Tarnished Plant Bug W 0.3"

Chinch Bug W 0.2"

AQUATIC BUGS

AQUATIC BUGS These are found in nearly every pond and stream. A few species are marine also. All are remarkable for their adaptations to life on or below the surface. Nearly all are predaceous, attacking other insects, snails, small fish, etc. In turn, these insects are food for longer fish and water birds. The water striders, taking advantage of surface tension on water, stay on the surface without breaking through. They skate along with remarkable speed. Other water bugs spend most of their time under water. Some carry down air on the surface of their bodies and use this for respiration. Other species breathe air that is dissolved in the water. In most cases, the young resemble the adults and mature after a series of nymphal stages. The water boatmen have an erratic swimming pattern. The backswimmers, as their name indicates, swim on their backs, but can also fly. The giant water bugs, sometimes 2 inches or more in length, prefer quiet water. Since their bite can produce a painful swelling, the amateur collector should exercise caution. When abundant these giant water bugs are harmful in fish hatcheries.

WATER SCORPION

These water bugs are elongated brown in colour and predaceous and have a long respiratory siphon emerging from the terminal end of the abdomen.

These appear like scorpions and feed on the small water dwelling insects. They have long posterior elongate siphon for respiration.

WATER SCORPIONS 3"

WATER BUGS

These are aquatic predaceous bugs. One of the important features of these bugs is that the female glues her eggs on the back of the male until hatching. and show parental care.

WATER BUGS 1.5"

DRAGONFLIES AND DAMSELFLIES These are seen near ponds and moist meadows. The former, also known as darning needles or stingers, are reputed to be dangerous. They are, but only to small insects which they eat on the wing. Dragonflies, larger than damselflies, rest with wings outstretched. The smaller, more delicate damselfly rests with wings folded. Both lay eggs on water plants or in water, and the nymphs developed there. These leave the water after several growing stages; the skin splits and the adult emerges.

MAYFLIES AND STONEFLIES These are unrelated yet are similarly adapted to aquatic environments. The 100 or more species of mayflies have transparent, veined wings and a long, forked tail. Myriads of short lived adults are seen on mating flights or near lights. The stoneflies include more than 200 species. The nymphs, like those of mayflies, live in water and are important food of fresh water fish. Both mayfly and stonefly nymphs take several years to reach the adult stage. Adult stoneflies have transparent wings, but do not fly very much or very often.

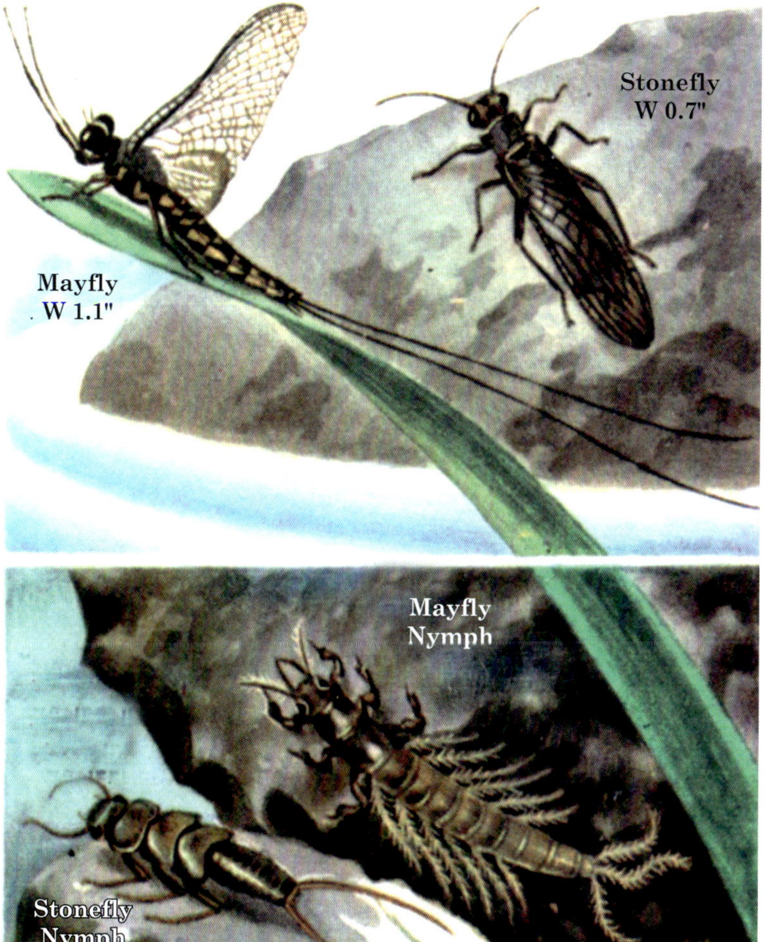

LACEWINGS These serve a double natural function. The adults are sometimes eaten by birds. The larvae feed on aphids and other destructive insects, earning the name aphidlions. Many of the larvae of the 20 or more kinds of brown lacewings cover themselves with remains of aphids and other debris. Golden eye lacewings lay stalked eggs which the larvae sometimes eat. Larvae spin silky cocoons from which they emerge as delicate, thin winged adults.

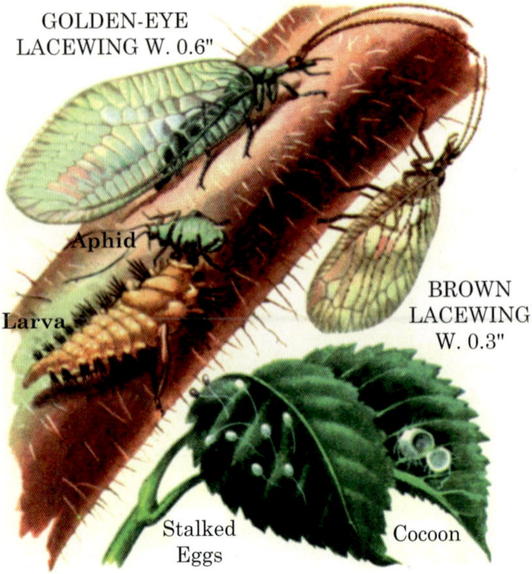

DOBSON FLY The ferocious looking adult male is harmless. The long mandibles are used in the mating, which ends its short life. The female lacks these exaggerated mouth parts. She lays a mass of thousands of eggs on plants overhanging a pond or stream. The larvae emerge, drop into the water, and spend the next three years feeding on smaller water life.

ANT LIONS The ant lions are so named because the larvae of members of this family have odd feeding habits. Eggs are laid on the ground. When one hatches, the larva digs a pit in sand or sandy soil and lives almost completely buried at the bottom. Should an ant or other small insect tumble in, it is seized in powerful jaws and sucked dry. As the larva matures, it builds a silken cocoon, in which it pupates. The adult resembles a miniature, drab damsefly with short antennae.

SCORPIONFLIES These flies are not poisonous, and resemble scorpions only in the modified tips of their abdomens. Eggs are deposited in the soil, where they develop into larvae. After the larvae pupate, the fly like adults emerge to live as scavengers, feeding on dead or disabled insects. Adults are found on plants. They do not fly well or often.

CADDISFLIES These include 17 families and over 200 species. The unusual larvae of caddisflies, so common in fresh water, are best known. These build cases of sand or plant debris, cemented together by silk. Cases built by different species are distinctive. After the adults mate in flights over the water, the female lays several hundred eggs on submerged rocks or plants. Larvae feed on small water plants and animals, and in turn are food of many fish. Larvae pupate in the larval case.

Adult W 0.9"

Larvae

Caddis Fly

BUTTERFLIES AND MOTHS

Antennae of butterfly (left) and moth

SCALES OF BUTTERFLIES

The largest, most attractive, and best known insects are grouped together in the order Lepidoptera, the butterflies and moths. All, except very few, have two pairs of wings. These and the body are covered with scales which give moths and butterflies their colour. The mouth-parts of adults are modified into a sucking tube which is rolled int a tight coil when not in use.

SCALES OF MOTH

Lepidoptera have four stages of development; egg, larva (caterpillar), pupa (cocoon or chrysalis), and adult. Most butterfly eggs are laid singly or a few at a time and are unprotected. Many moths lay a large number of eggs in one place and may cover the egg mass with a protective coating which includes hairs and scales from the female's body. Both the caterpillars of moth and butterflies have chewing mouth parts and some kinds do tremendous damage to crops. Most caterpillars have 6 true legs on the thorax, and from 4 to 10 unjointed false legs on the abdomen. A few have irritating hair or spines. Many caterpillars spin a silken cocoon, sometimes covered with hairs, in which they pupate. The butterfly larvae make no cocoon but form a less protected chrysalis. Some Lepidoptera winter as pupae; some emerge in a short time.

Giant Swallowtail Caterpillar

The adult moth and butterfly are usually quite different, though one group, the skippers, shows intermediate characteristics. Butterflies usually fly by day; moths ordinarily fly by night. The former customarily rest with their wings folded back; moths rest with their wings in a horizontal position. The antennae of butterflies are thin, ending in a knob. Those of moths never end in knobs and are often feathery. Though some butterflies are more attractive, the moths form a larger, more diverse, and more important group. Butterflies and moths are not only beautiful but have other characteristics too. For example, the *Cethosia mytina* is not only a very beautiful butterfly but it has a sweet perfume which persists even when dead.

PUPA

COCOON

MONARCH This common, attractive butterfly has spread into Asia, America, Africa and Australia. Males are identified by a black spot on the third vein of the hindwing. Two or three generations grow in one summer. In fall, swarms of adults migrate southward, covering entire trees when they stop to rest.

MONARCH 1.3"

Larva on milkweed

Chrysalis

LARVA ON POPLAR

VICEROY
W 2.8"

They do not hibernate, as is sometimes believed. The pale green, conical eggs hatch in 3 to 5 days into larvae which feed on milkweed. The larvae are not often eaten by birds.

VICEROY This butterfly is noted because in colour, pattern, flight, and habits it mimics the monarch. Like the monarch it is abundant in late summer. But the viceroy is smaller and has curved black lines crossing the veins of the hindwings. Its eggs, and the larvae which feed on poplar and willow, resemble those of the purples to which the viceroy is related. The larvae hibernate in rolled leaves for the winter. There are two generations or more a year.

PURPLES The banded purple has a conspicuous white band across its wings with a border of red and blue spots on the hindwings. Eggs are laid on leaves of willow, birch, and poplar, on which the larvae feed. The red spotted purple is so named for its red spots on the under side along the wing borders and at the base of the hindwings. The larvae feed on wild cherry, willow, and other trees, preferring shaded woods.

BANDED
PURPLE
W. 2.9"

RED-SPOTTED
PURPLE
W. 3.1"

BUCKEYE It is most common in West America. The eye spots make identification of the butterfly easy. The buckeye lays its green, ribbed eggs on plantain, stone crops, or gerardia, and it is on these plants that its larvae feed. The larva forms a brown chrysalis from which the adult eventually emerges. Buckeyes are medium sized butterflies. They are often found in open fields.

BUCKEYE
W 2.5"

FRITILLARIES The fritillaries, one of the largest groups of butterflies, are cosmopolitan. The family of which they are the most important members has front legs which are reduced in size and held close to the body. Only 4 of the 6 legs are used in walking. Fritillaries are mostly medium sized butterflies, orange or reddish above, with silvery or light spots on the under side of the hindwings. Sometimes the males are a brighter red on top than the females.

The eggs are generally conical in shape and ornamented with ridges. All the caterpillars are spiny, with the spines on the head a bit longer than the others. Most feed at night on such plants as violets, goldenrods, and other composites. While the early stages of the common fritillaries are known, there are many less common species. and information on the egg and caterpillar of these species is incomplete. The chrysalis of fritillaries is usually angular, forked at the top, and bordered with knobs. It is often brownish.

Great Spangled Fritillary W. 3.5"

Meadow Fritillary W. 1.7"

Variegated Fritillary Larva

Upper side Regal Fritillary W. 3.4"

Silver Bordered Fritillary W. 1.5"

Gulf Fritillary W. 2.9"

Gulf Fritillary Larva

Upper side

REGAL FRITILLARY W. 3.4"

Great Spangled Fritillary Larva

CHECKERSPOTS AND BALTIMORES. Of this group of small, attractive butterflies, the two illustrated here are well known. The Baltimore caterpillar is black, with orange bands and black bristles. It feeds on turtlehead and related plants. Baltimores are common, often in moist areas. The checkerspot lacks the brilliant orange of the Baltimore. The caterpillar is entirely black, with longer bristles. It feeds on painted cups and monkey flowers, which are common in the West America.

CHECKERSPOT
W. 1.6"

BALTIMORE
W. 1.9"

Turtlehead

QUESTION MOUTH W. 2.5"

COMMA W. 2"

ANGLE WINGS This group of small and medium sized butterflies has angular, notched forewings. The hindwings often have short tails. The question mark (also called the violet tip), comma, and the mouring cloak are best known. The brownish larvae of the question mark feed on hop, elm, and nettle. The greenish often pale comma caterfillar feeds on nettles. Both caterpillars have branched spines on each body segment.

MOURNING CLOCK This butterfly, like the other angle wings, hibernates as an adult and hence makes its appearance very early in spring. Dark, barrel shaped eggs are soon laid on twigs of poplar, elm, hackberry, or willow. The gregarious caterpillars occasionally injure a tree by stripping the foliage. The mourning cloak is common and widely distributed over the entire northern hemispheres.

American Painted Lady
(Hunter's Butterfly)
W.2.0"

Red Admiral
W.2.0"

Larva on nettle

RED ADMIRAL, AMERICAN PAINTED LADY, and PAINTED LADY Two of these three closely related butterflies are difficult to distinguish. But the red admiral is clear and unmistakable because of the red bands on its forewings and red borders on its hindwings. The red admiral is found the world around in the northern hemisphere. The light green eggs are barrel shaped. The caterpillar feeds on hops, ramle, and nettles. American painted lady is restricted to North America. Its upper side is very similar to that of the painted lady. The under side is distinct in having two large ``eye spots" which the painted lady lacks. The American's larvae are spiny and black, with rows of white spots. It feeds on arrow weed, cudweed, and other everlasting. The painted lady or thistle butterfly is reported to be the most widely distributed of all known butterflies. This may be because burdock, thistles, sunflowers, nettles, and other plants which larvae eat are widely distributed also. Caterpillar is greenish, with black spots and light, branched spines.

PAINTED LADY
W. 2.2"

NYMPHS AND SATYRS More than 60 species make up this large group of small to medium sized butterflies, most of which are dull brownish or gray in colour. Their characteristic markings are eye spots on the under sides of the wings. Most prefer open woods and mountain areas. The caterpillars of these species are smooth, and taper towards both the head and the tail, which is always forked. The larvae of the eyed brown have a pair of red horns at each end. Their food is grass and sedge. The pearly eye is slightly larger in size, and in larvae similar in appearance. The common wood nymph has a pair of yellow patches on each forewing surrounding the purplish eye spots. The male has two small eye spots on the hindwings. Markings vary a good deal. The larvae lack the horns of the first two species. The little wood satyr is smaller, with eye spots on both fore and hindwings. It prefers open forests and thickets. The larva has no red at all, its colour varies from greenish to pale brown. It, too, feeds on grasses.

COPPERS AND BLUES This large family of small butterflies includes over 2,000 species. These butterflies are common. The family has three groups. The hair streaks are usually brownish or bluish, with hair like tails at the tip of the hindwings. The second group, the coppers, are nearly all a copper red colour with black marking. The last group, the blues, are very small. Western pygmy blue, with a wingspread of a bit over 0.5 inch, is our smallest butterfly. Blues are more common. The caterpillars of all this family are short, thick, and slug like. Some are flattened and covered with fine hair. They feed on a variety of plants including oak, hickory, hops and sorrel. One species has a carnivorous caterpillar which devours plant lice.

COPPER BUTTERFLIES

EASTERN TAILED BLUE
W 1.0"

SPRING AZURE
W 1.1"

under

upper

MARINE BLUE
W 1.1"

WESTERN PYGMY BLUE
W 0.6"

BLUE BUTTERFLIES

CABBAGE BUTTERFLY The cabbage butterfly is a serious pest of almost all the vegetable plants of the family Cruciferae. It is world wide in distribution. The damage is caused only by the larvae which are known as "cabbage caterpillars". The butterflies are pale white with a smoky shade on the dorsal side of the body. The wings are pale white with a black patch on the apical angle of each fore wing and a black spot on the costal margin of each hind wing. The females have two distinct black circular dots on the dorsal side of each fore wing. The males are smaller than the females and have black spot on the lower side of each fore wing.

The butterflies are very active in the field and lay about 164, yellowish conical eggs, in clusters of 0750-90 on the upper or lower side of a leaf. The eggs hatch in about 2 weeks. The larvae are pale yellow when young and turn greenish yellow later on. The head is black and the whole body is marked with black spots. They pupate in barns or on the trees and the butterfly breeds 4 times in a year.

CABBAGE BUTTERFLY
W 1.8"

APOLLO BUTTERFLY
W 2.5"

PEACOCK BUTTERFLY
W 2.0"

OWL BUTTERFLY W 6-8"

OWL BUTTERFLY

The huge underside eye spots of owl butterfly gives this species if name when its wings are inset position and held as shown here, head downward its likeness to the head of an owl is very striking. The butterfly holds its wings quite differentially when resting in a natural position and it is not imitating an owl; when disturbed it suddenly flinches, flapping its great wings those terrifying eye spots are enough to startle any predator giving the butterfly time to scape, out butterflies are found only in S. America.

CETHOSIA MYTINA
W 3.0"

Hypolimnas misippus W 2.5" (Male)

Hypolimnas misippus W 2.5" (Female)

MORPHO BUTTERFLIES

Morpho butterflies occur in South tropical America and North India are fairly large, have brilliant metallic colour with iridescent blue wings. There are several species of morpho butterflies. Morpho larvae are hairy and they often live gregariously.

Morpho rhelenor helena W 6-7"

Stichopthalma camdeva W 5"

STICHOPTHALMA CAMDEVA

It is a very large round winged species which is found chiefly in the northern India. It has slow flapping flight and is out at dawn and dusk.

Polyura delphis
W 3"

Arhopala centaurus
W 1.5"

Papilio paris W 3-4"

Delias eucharis W 2.5"

Eurema hecabe W 1"

Danaus genutia W 2.5"

Ixias pyrene W 1-2"

Delias hyparete W 2.5"

SLEEPY ORANGE W 2"

Bibasis gomatae W 1"

Melanitis leda
W 2-3"

MELANITIS LEDA

It is a fairly, large satyrid butterfly. It is one of the most widely distributed species in Asia. It extends beyond Asia and undoubtedfly occurs in all parts of the continent. This species is extremely variable in size, pattern and colouring. It flies at dusk as well as by day.

Doleschellia bisaltidae W 2"

Kaniska canaee W 2"

PAPILIO SP. (BUDDHAS PECOCK) W 2.5"

Papilio harishanus W 4-5"

Limnas chrysippus W 2-3"

Precis orithya W 1.5"

Vindula arsinae W 2"

Limnas alcippus W 2-3"

Thaumantis diores W 4"

Kallima philarchus W 4"

GIANT ORANGE TIP W 3.5"

Melanargia sp. W 2.5"

Hestia leuconoe W 3"

Papilio sarpedon W 2-3"

Papilio rumonzovia W 3"

Teinopalpus imperialis W 3"

Papilio hector W 3"

Papilus kotzcbuea W 2.5"

LACEWING W 3.5"

Lacewing butterflies, The species of *Cethosia* occur from India through Malaysia and Indonesia. Hongkong to northern Australia.

Papilio polymnestor W 2.5"

Symprenthia hippoclus W 2"

Delias aglaia W 2"

COMMON ROSE W 3"

PHALANTHA PHALANTHA W 3"

ANAR BUTTERFLY W 1"

ANAR BUTTERFLY

The anar butterfly is distributed throughout India and is a serious pest of pomegranate. Besides anar, the list of food plants includes guava, ber, tamarind, apple, orange etc. Damage is caused by the larva which bores into the fruit and feeds on its contents. From the wound caused by the larva, offensive smelling matter oozes out as a result of which the fruits are infected with a number of bacteria and fungi.

UNION JACK

In Australia, New Guinea and Queensland are found almost all the worlds species of *Delias*. The above shown is *Delias mysis,* a tropical pierid.

UNION JACK BUTTERFLY W 2"

Papilio polytes romulus W 2.5"

Bhutan glori W 3-4"

Euploea callithae hansemanni W 3"

Papilio childrene oedippus W 3"

The male papilio childrenae oedippus has white scent scales on the open marginal folds of his hind wing.

Euploea diocletianus W 3"

Euploea mulciber W 3"

Cyrestris thyodamas W 2"

Neptis sp. W 1.5"

PANSY W 1"

W 2"

*Precis octavia (*Dry season) *Precis octavia* (Wet season)

Lampides brohetics W 1"

BLUE LEAF BUTTERFLY W 3"

Apatura ambica W 2"

INDIAN LEAF BUTTERFLY W 2.5-3"

INDIAN LEAF BUTTERFLY

In these butterflies the wings. are exactly shape of leaves and tail looks like stalk marks on underside closely resemble the midrib and veins. Some are like the dry leaves and marked with spots like leaf decay. The markings and ground colour are very variable. No two butterflies are exactly of the some shade and colour as others.

Srilanka has developed many species of its own for e.g. blue leaf butterfly *Kallima paralekta* and, *Philarchus*.

Troides aeacus (male) W 4-7"

Troides aeacus (female) W 4-7"

TRIODES ACACUS

Troides acacus is one of the commonest of a group of large black and yellow bird wings and comes from China, India and the Malay Peninsula. The Larvae feed on *Aristolochia*.

Leptocirus curius W 5.5"

Apatura sp. W 1.5"

Papilio memnon agenor
(tailed from) W 4"

Papilio memnon heronus
(tailed from) W 4"

Papilio memnon
(tail less from) W 4"

Papilio blumei W 5"

Papileo coon W 3-4"

Trogonoptera brookiana W 6-7"

Prothoe calydonia W 3.5"

Elymnias hypermnestra W 2-2.5"

Rhopalocampta benjamini W 1.5"

Lycaena sonorensis W 1"

Doritis apollinus W 2.5"

Neocheritra amrita W 1.5"

Eupolea core W 2.5"

Ancyluris formosissima W 1.5"

Arhopala centurus W 1.5"

Jamedes sp. W 1"

Papilio arcturus W 2.5"

Papilio alexanor W 2-3"

LEMON BUTTERFLY
W 2.5"

LEMON BUTTERFLY

The butterfly is beautiful, measuring from 2.5" wing tip to wing tip. The head and thorax of the butterfly are black; and the underside of the abdomen is creamy yellow in colour, the wings being dull black which are ornamented with yellow markings. There are 3-4 generations in a year and the winter is passed in the chrysalis stage.

This is not only a pest of citrus but of all the plants belonging to family Rutaceae. The young larvae feed on the tender leaves while the older ones on the mature leaves. As a result of heavy infestation the entire plant may be left without leaves and bears no fruits.

SULPHURS Dozens of common or clouded sulphurs hover as a yellow cloud over roadside puddles. Their colour is variable, females are paler, with yellow spots in borders of forewings. There are two generations annually. The alfalfa butterfly, also varying in colour, is sometimes pale. It also is common along roadsides, in fields and gardens. The larva is similar to the common sulphurs but has pink stripes. Both these larvae feed on clover, alfalfa and other legumes.

SPICEBUSH
SWALLOWTAIL
W 3.5"
PARASSIUS
W 2.4"
GIANT
SWALLOWTAIL
W 4.4"

ZEBRA
SWALLOWTAIL
W 2.7"
BLACK
SWALLOWTAIL
W 2.8"
TIGER
SWALLOWTAIL
W 3.8"

SWALLOWTAILS The swallow tail butterfilies are found in America, Asia, Africa making the swallowtails a group that is widely known, admired collected, and studied. Closely related to the swallowtails are the parnassians, more common in the America and Asia. These lack the `tail' on the hindwings that gives the swallowtails their name. The caterpillar pupate on the ground. The mountain butterfly is an example of this group.

The swallowtails are predominately black or yellow. Some species occur in several forms. The female tiger swallowtail may be either yellow or black, the black form being more common in the South America. Swallowtail eggs are usually round, fattened at the base. The caterpillars are generally smooth, lacking spines, though the pipevine swallowtails have fleshy horns. All the larvae can emit a rather strong, musky odour which may protect them. The chrysalis rests on its tail end, supported by a loop of silk of the middle. There are often two broods a year.

The swallowtails illustrated here on various pages are easy to identify in spite of variations in colour. Their caterpillars and food plants are shown on the next page. The zebra swallowtail has the longest tail of any native species. The giant swallotail is the largest, with a spread of 4 to 5 1/2 inches. Its larvae are occasionally destructive to citrus orchards. Both male and female black swallowtail have a double raw of yellow spots, but those of the female are smaller. Yellow spots on the forewings and greenish hindwings mark the spicebush swallowtail. The pipevine swallowtail lacks the yellow spots.

SKIPPERS About 200 kinds of skippers are found in America and this is only one tenth of the total number. Their rapid, darting flight gives them their name. These small butterflies have characteristics of moths. Some rest with the hindwings or both wings horizontal as moths do. The smooth caterpillars have large heads and thin ``necks''. They feed on locust, clover, and other plants. The silver spotted skipper is the most common of the large skippers. Many are much smaller than this. A large number of skipper are found in Asia.

SPICEBUSH on Sassafras

ZEBRA on pawpaw

PIPEVINE on pipevine

BLACK on Parsley

GIANT on orange

TIGER on wild cherry

SWALLOWTAIL CATERPILLARS
and their food plants

SKIPPER BUTTERFLIES

ACREA MOTHS and BANDED WOOLLYBEARS Acrea is one of the most common moths, easily identified by the spotted abdomen. The male is illustrated. The female's hindwings are white. Caterpillars feed on grasses and garden plants. The banded woolly bear or isabella moth caterpillar abundant in autumn, feeds on plantain and related plants. It hibernates as larva and pupates in spring. Both larvae and pupae use hair in making cocoons.

BANDED WOOLYBEAR W. 1.8"

SPHINX MOTHS These moths are common in America and Asia. Their common names, as tomato worm and tobacco worm, indicate the foods sought by some larvae. Leaves, and sometimes fruits, are eaten. Some larvae feed on potatoes and on wild members of the potato family. Other species eat birch, willow, grape, and other plants, occasionally damaging nurseries and vineyards. The larvae are large and usually have a tail or horn. Some rear back into a belligerent attitude when molested, but none are poisonous, as is sometimes believed. Braconid wasps lay their eggs in the living caterpillars, which the wasp larvae eat. Caterpillars of Sphinx moths covered with wasp cocoons are often seen. Caterpillars pupate in the ground. The common adults are identified by abdominal or wing markings. The sucking-tube mouth is long, enabling Sphinx moths to get nectar and pollinate tubular flowers, such as nicotina, petunia, honeysuckle and trumpet vine.

SPHINX MOTHS

AILANTHUS SILK MOTH
W. 4.2"

AILANTHUS SILK MOTH This moth was imported from China to America. Where a coarse grade of silk is obtained from its cocoons. The large caterpillars are controlled by natural enemies, and since they feed chiefly on ailanthus, a weed tree, they are not harmful. These are the only large moths with white tufts on the abdomen.

CECROPIA MOTH
W. 5.5"

CECROPIA MOTH The large, tubercle-studded cecropia larvae feed on cherry, maple, willow, and many other plants. The large, tough, brown cocoons are firmly attached to branches and are easily found in winter. Outdoors, the huge moths emerge in late spring or summer, but when cocoons are brought indoors they hatch earlier. The emergence of the adult is a sight to see. The wrinkled, velvety wings unfold till they are 5 or 6 inches across-something worth watching.

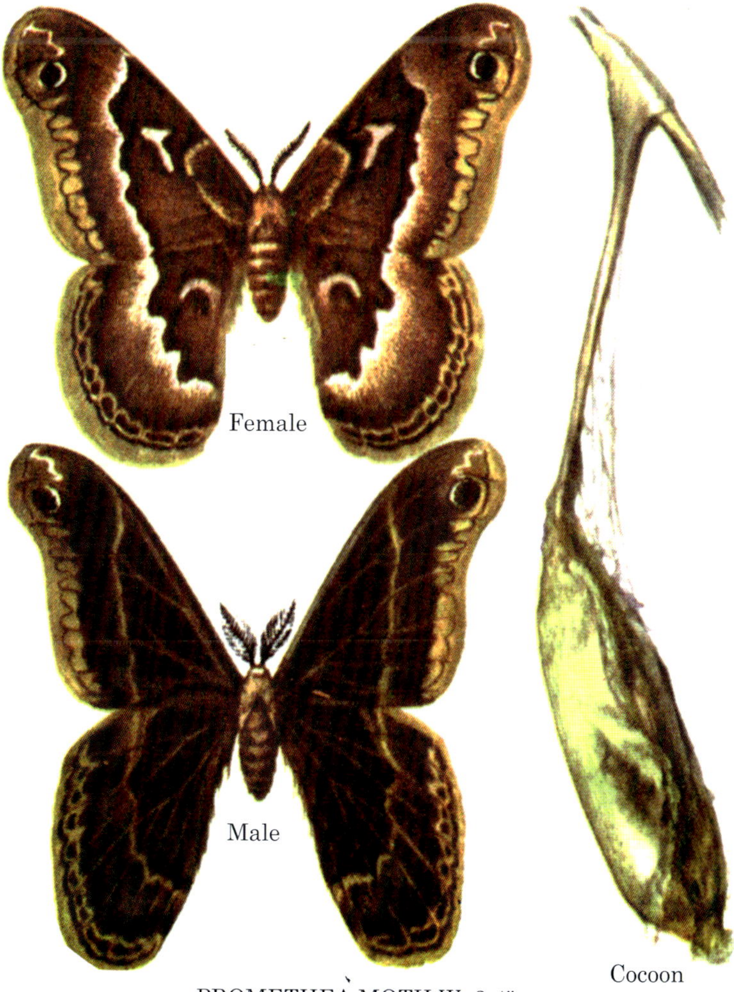

PROMETHEA MOTH W. 3.4"

PROMETHEA MOTH This moth, has a bluish green larva with two pairs of short red horns near the head. It feeds on wild cherry, tulip tree, and sweet gum, as well as on spicebush. Collect the compact cocoons, each wrapped in a dry leaf, and watch the adults emerge in spring. Recognize the male by its darker maroon colour. The female is a bit larger, lighter, and browner, with slightly different markings.

POLYPHEMUS MOTH W. 5.3"

POLYPHEMUS MOTH Because of its gigantic size and the ``eye-spots" on the wings, this night flying silk moth was named after the one eyed giant, Polyphemus of Greek mythology. The green larvae, sometimes over 3 inches long, feed on oak, hickory, elm, maple, birch, and other trees and shrubs. They spin their plump cocoons either on the ground or attached to a twing.

IO MOTH W. 2.5"

IO MOTHS Three very similar species are found in America. The larvae are of particular interest, for their sharp spines are mildly poisonous. Recognize them by their horizontal red and white stripe. Handle them with care when you find them on corn and other garden and wild plants. Ichneumon flies often attack the larvae. The cocoon is found on the ground in dead leaves. The adult male is smaller than the female and has bright yellow forewings.

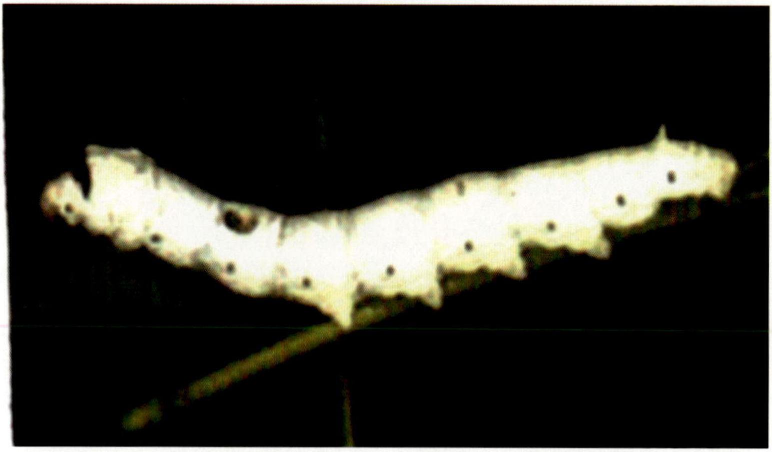

Caterpillar

MULBERY SILK MOTH W. 2.5"

MULBERRY SILK MOTH The silk produced by mulberry silkworm is supposed to be superior in quality to its other types due to its shining and creamy white colour. The caterpillars feed on the mulberry leaves. It has been reported that mulberry silk worm prepares silk filaments several thousands meters in length at the rate of 15 cm per minutes.

SUNNHEMP MOTH W. 1.5"

SUNNHEMP MOTH The caterpillar is known as ``sunnhemp hairy caterpillar". The caterpillar feeds on the leaves and capsules of sunnhemp. The adult moth is creamy white with black and red spots on the fore wings and dark black spots on the margin of the hind wings. The full grown larva measures about 3.8 mm in length and is provided with red, white markings on the body, the head is slightly brown in colour.

RED UNDER WINGS MOTH The red underwing moths are widly distributed. The brightly coloured underwings of the moth are concealed while at rest but on taking flight they are exposed and show a bright flash of colour.

RED UNDER WINGS MOTH W. 1.5"

TASSAR SILK MOTH W. 10"

TASSAR SILK MOTH The Tassar silk moth is also extremely beautiful like other silk moths. The caterpillars (Larvae) feed on the leaves of Arjun, Sal, Ber, Jamun etc. The tassar silk is coppery in colour.

GIANT SILK MOTH W. 12"

GIANT SILK MOTH This moth is found from India East Worlds through China to Malaysia. It is reputed to be the largest moth in the world. The marking on the wing tips are reminiscence of snakes head. The larvae feed on many trees like willow rhododendron but they thrive well on the tree of heaven.

ERI SILK MOTH W. 6"

ERI SILK MOTH The Eri Silk moth bears same pattering as the giant silk moth. It is about half the size (w. 6") of giant silk moth. Their larvae are also white and similar in appearance. The larvae feed on Eri-leaves hence the name Erisilk worm. The colour of silk is creamy white but is less shining than the mulberry silk. The cocoon has loose texture and the silk is not reliable and is of poor quality.

DIAMOND BACK MOTH This insect is world wide in distribution and is a serious pest of cauliflower and cabbage. It also feeds on other cruciferous, solanaceous and lillaceous plants. The damage is cased by the caterpillars which in earlier stages feed in mines on the under side of the cabbage leaves and in the later stages feed on exposed leaves.

DIAMOND BACK MOTH W. 1"

INDIAN MOON MOTH W. 5"

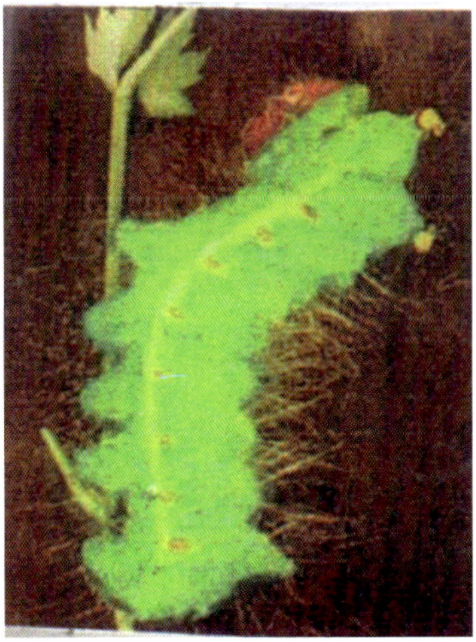

Larva

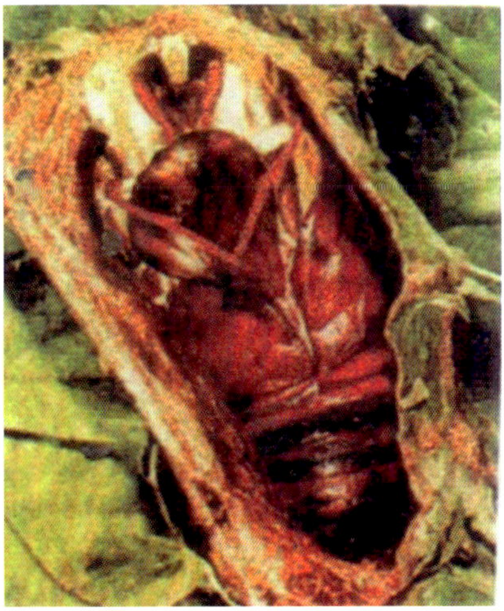

Cocoon cut open to show pupa

INDIAN MOON MOTH Spectocular among all moths is the Indian in moon. The male is beautifully kite shaped and coloured in softest shades of green yellow and maroon. The female has lovely furl tails. The moths breed easily in captivity. Their larvae are coloured with patches of red and black at first then they become red all over with black spots and at the third skin change they are green, covers with orange tubercles. The larvae feed on rhododendron, apples and other common trees.

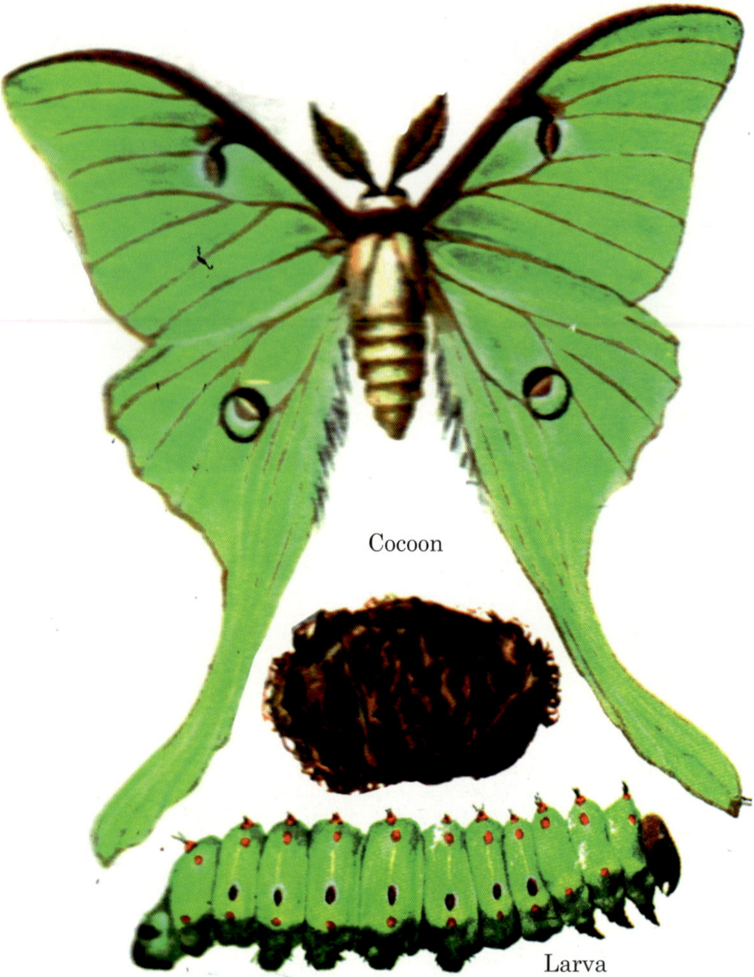

LUNA MOTH W. 3.2"

LUNA MOTH This handsome moth, with its striking long tails and delicate green colour, makes a lasting impression on those who see it for the first time. It is a favourite with collectors of moths and butterflies. The larvae, smaller than the other night fliers, feed on sweet gum, walnut, hickory, and persimmon. The cocoons are usually spun on the ground. Male and female are similar in appearance. Some have more purple on the borders of the wings than others.

Clouded locust underwing

UNDERWING MOTHS

UNDERWING MOTHS There are over 1000 species of these attractive moths in the United States and a variety of underwings can be found in every locality. When the adult rests on bark with its wings folded, it can scarcely be seen. In flight, the bright colours of the under wings are in sharp contrast to the drab pattern of the forewings. Found in America, Asia and Africa.

IMPERIAL MOTH W. 4.6"

Larva on white aak

IMPERIAL AND REGAL MOTHS These moths are closely related to the large silk moths. The hairy imperial caterpillar, with short horns near the head, varies from green to brown. It feeds on pine, hickroy, oak, maple, and other trees. No cocoon is formed, the pupa resting in the ground. The forewings of the male are purplish those of the female are richer in yellows. The larvae of regal moths, which feed on walnuts, have large, red, curved horns.

CORN EARWORM AND BORERS

CORN EARWORM AND BORERS Everyone who has husked sweet corn has seen the greenish or brown larva of the corn earworm, found almost everywhere. It feeds on other garden crops, too, and pupates underground. The larvae bore into stalks, weakening and breaking them. Control is difficult, as borers live in wild plants also.

WHITE-MARKED TUSSOCK MOTH
and larva on maple

WHITE MARKED TUSSOCK MOTH The larvae, often seen in late summer, are recognized by their tufted white hairs and attractive contrasting colours. They are the pests of most shade and ornamental tree and are best controlled by destroying cocoons in winter. The egg masses are laid by the wingless females on the surface of their cocoons. Eggs hatch in spring, but the larvae are not noticeable till summer. Adults are small and inconspicuous.

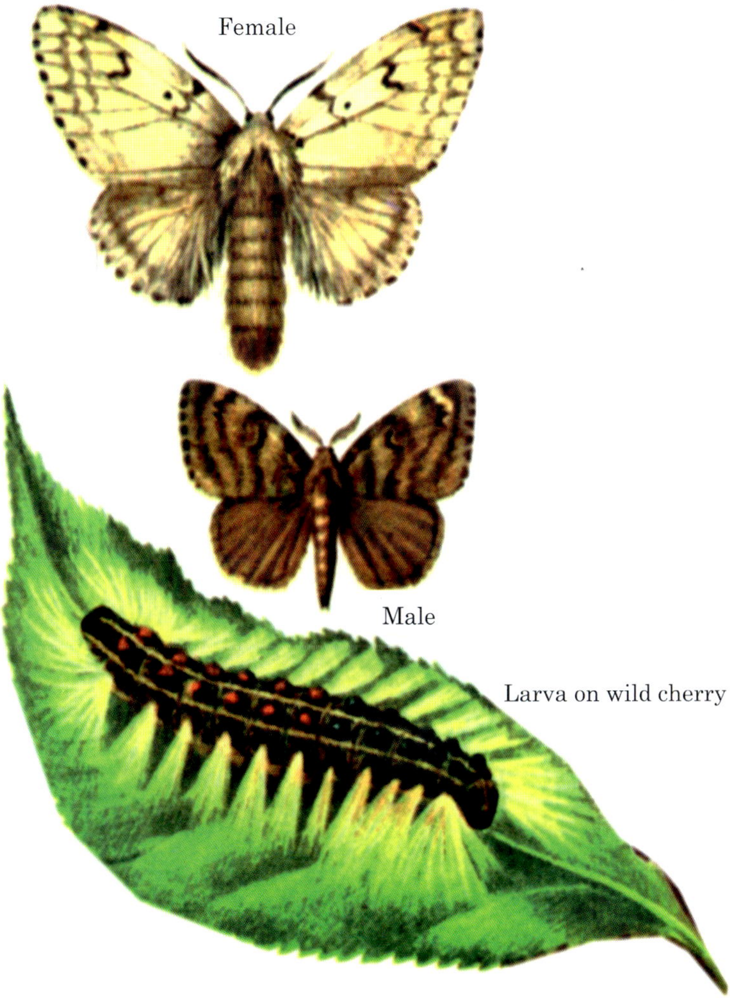

GYPSY MOTH W. 1.9

GYPSY MOTH The gypsy and the brown tail moth are European relatives of the white marked tussock moth. The larvae eat the leaves of most shade and forest trees and they tend to feed at night. Larvae pupate in midsummer and adults emerge shortly. The female lays a mass of eggs, covering them with hair and scales. These hatch in spring.

FALL WEBWORM
W. 1.5"

TENT CATERPILLAR
W. 1.2"

TENT CATERPILLARS AND FALL WEBWORMS These two pests of forest, shade, and ornamental trees are not closely related. Both build webs, but that of the fall webworm covers the leaves. The tent caterpillar web, built only in spring, is usually at the crotch of branches. Adult webworms, with variable black spots on forewings, lay eggs on leaves. Tent caterpillar egg masses are found on twigs.

FALL CANKERWORM
W. 1.2"

CANKERWORMS These pests of apple, other fruit, and shade trees are worth watching as they lope along on their true and false legs. Sometimes they spin a silken thread and hang suspended in mid air. Fall and spring cankerworms are equally abnoxious. The wingless females lay their eggs on bark. After feasing on leaves, the larvae pupate underground. The adults emerge late in autumn and can be trapped by bands of sticky paper around tree trunks.

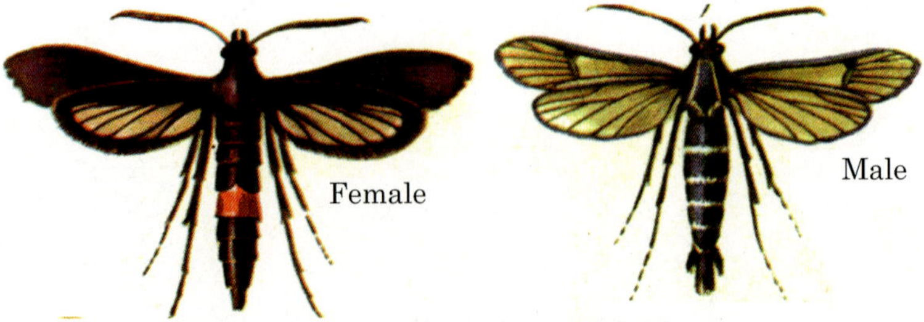

PEACHTREE BORER W. 1.1"

PEACHTREE BORER The peachtree borer larvae enter the tree as soon as they hatch from the eggs, which have been laid on the bark. The larvae burrows can be recognized by exuding gum on the surface. After spending the winter as a larva, the borer pupates in a crude cocoon. The adult emerges in about a month and mates. The females lays a new mass of several hundred eggs.

CODLING MOTH This pest is so widespread that spraying apple trees for its control is a routine operation. Tiny eggs are deposited on leaves. As they hatch, the larvae enter the new fruit. Later, larvae hatch from eggs laid on the apples. After feeding in the fruit, larvae pupate on the bark. There are often several generations in a year. Late larvae hibernate and pupate the following spring. This moth is cosmopolitan in distribution.

CODLING MOTH W. 0.8"

Larva
in bag

Male

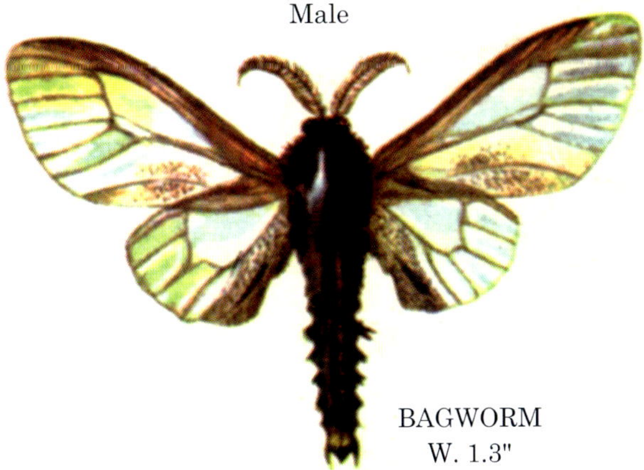

BAGWORM
W. 1.3"

BAG WORMS The wingless and footless female, after mating, crawls back into her ``bag" and lays hundreds of yellow eggs, which hatch in spring. The young larvae feed on leaves of many kinds of trees, building their conical bags as they feed. Later they bind their bags to twigs and pupate. The male emerges, seeks the female, and mates.

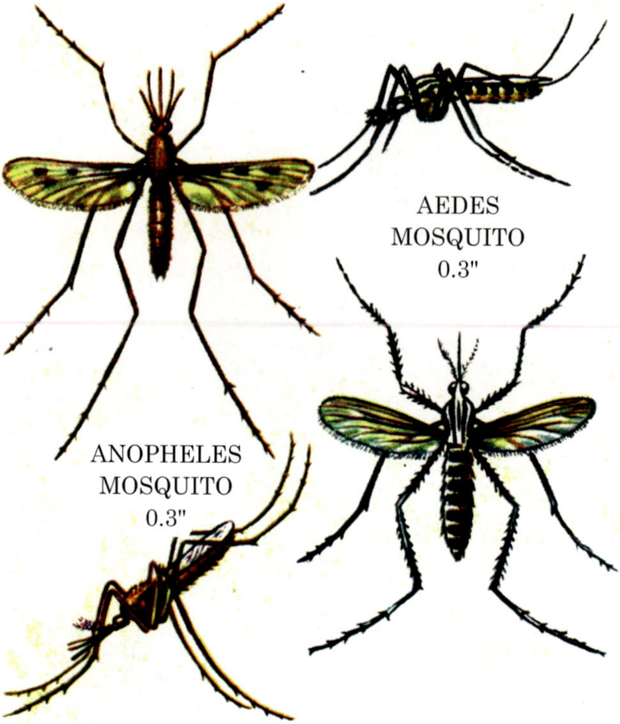

MOSQUITOES This large group of small but important insects has been amply studied as part of public health campaigns against malaria and yellow fever. The conquest of malaria is a scientific milestone. The common carriers of the disease, the anopheles mosquitoes, are recognized by the ``three pronged" beak of the female and by the tilted position they assume when resting. The aedes mosquitoes, one of which carries yellow fever, are more like the common house and swamp mosquito in appearance. The disease carrier is limited to tropics and subtropics. Female culex mosquitoes lay rafts of several hundred eggs, which hatch in a few days into larval wrigglers. In a week or so these pupate, and the adults soon emerge. Male mosquitoes are harmless. Their beaks are not fitted for piercing. Females bite and hence can transmit disease. They are the only ones which buzz. Mosquitoes have been controlled by draining swamps and by the wide use of insecticides.

HOUSE MOSQUITO

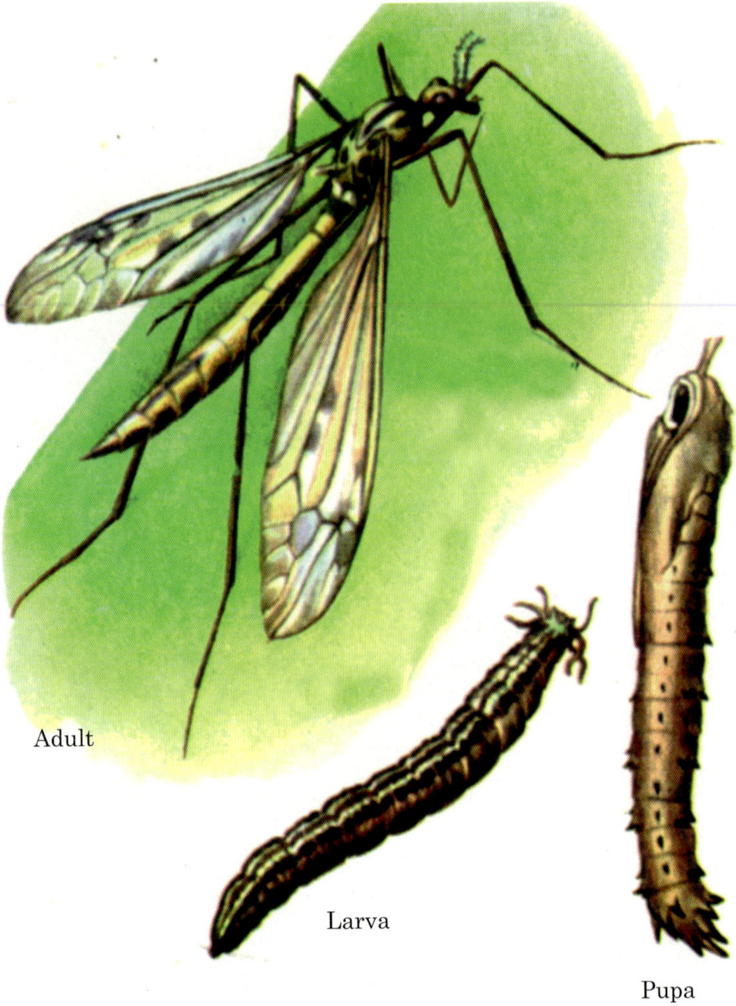

CRANE FLY W. 1.0"

CRANE FLIES Adult crane flies are sometimes mistaken for giant mosquitoes. But the adults, often seen around electric lights, may not eat at all. Some are apparently predaceous. The female lays several hundred eggs on damp soil. The larvae burrow into the ground or decaying wood. Only a few attack plants. In a few weeks they pupate, and in about a week appear as adults. These flies are cosmopolitan.

ROBER FLY W. 0.8"

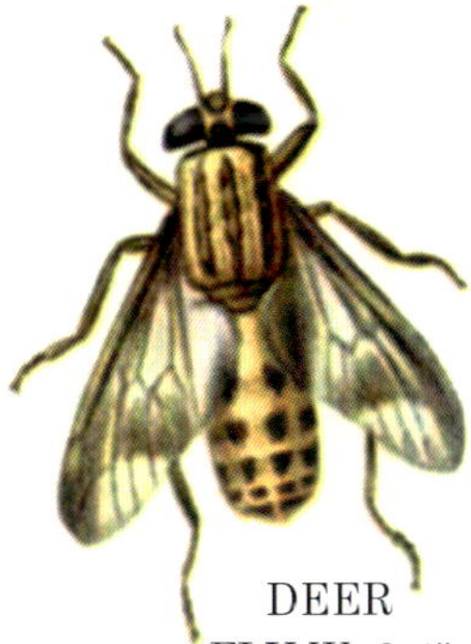

DEER FLY W. 0.4"

MARCH FLY W. 0.3"

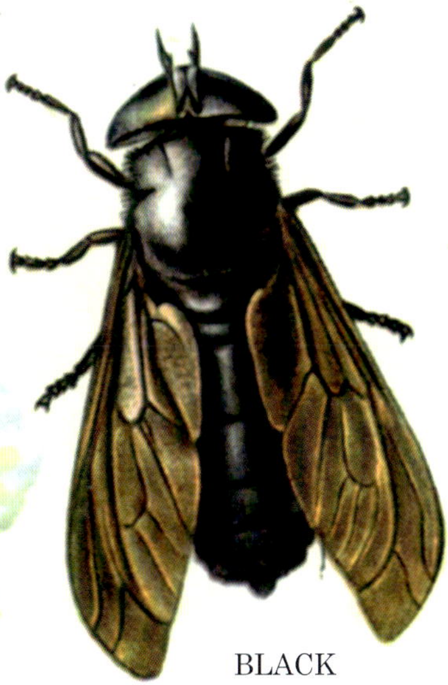

BLACK HORSE FLY W. 0.9"

FLIES The flies described here include serious pests of plants and animals which cause losses, running into millions. The vast majority are harmless; some species are beneficial. The robber fly preys on insects, some larger than itself. The deer fly and the black fly can make a camper miserable. March flies (which are more common late in spring) are often seen on flowers. The black horse fly, sometimes a full inch long, bites severely. Many flies transmit disease as they bit infected animals and then others. Insect repellents and insecticides are useful in controlling them.

Larva

BLUE BOTTLE
FLY W. 0.5"

GREENBOTTLE
FLY W. 0.5"

Pupa

BLUEBOTTLE AND GREENBOTTLE FLIES These flies are attractive insects, but thoroughly abnoxious otherwise. Eggs are laid on dead animals, garbage, sewage, or in open wounds of animals. Some related species parasitize and kill animals and even man. Eggs hatch very soon after being laid; larvae are mature in less than 2 weeks. The short life cycle means several generations a season; only continual spraying will keep these pests under control.

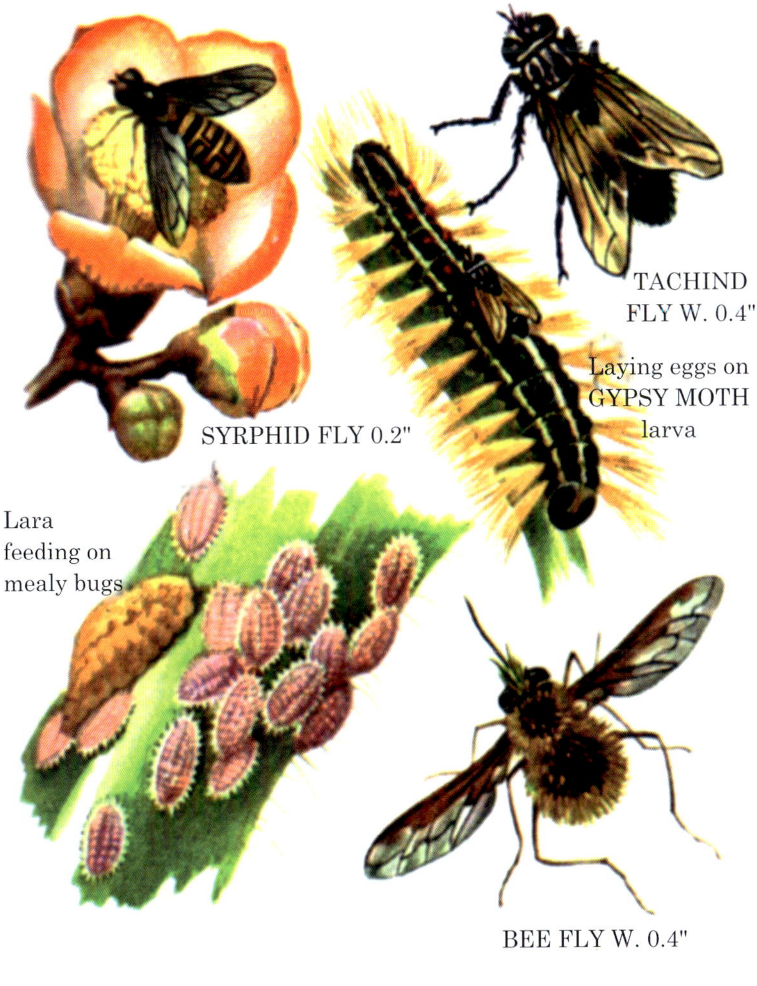

TACHINID AND OTHER FLIES Tachinid flies are beneficial insects which help control injurious ones. More than 1,400 species have been described. Because these flies are prolific, their value as parasites is increased. Syrphid flies, known as flower or drone flies, are similar insects. They are often seen approaching a flower, coming to an abrupt stop and hovering in mid air. The larvae eat aphids and scales. The bee fly lives in hives, and its larvae often attack and feed on larvae of bees and other insects.

Male W. 0.2"   Beaded wing   Vestigial-wing

FRUIT FLY These small and rather important flies are often seen around rotting or fermenting fruit. Their claim to fame rests on scientific uses to which they have been put. Fruit flies have been used in hundreds of experiments dealing with inheritance. They have probably been studied from this angle more intensely than any other animal. The fact that their life cycle is less than 2 weeks enhances their value in this work. They are easily grown in the laboratory, where interesting forms have appeared naturally or from exposure to experimental radiation.

SIX-SPOTTED TIGER BEETLE
W. 0.5"

PURPLE TIGER BEETLE
W. 0.5"

TIGER BEETLES These handsome beetles are often seen on summer afternoons darting in and out of paths. They are widely distributed and quite common, but agile, swift, and difficult to catch. Eggs are laid in the soil. The predatory larvae. dig deep burrows and wait at the openings to catch passing insects in their powerful jaws.

ROSE CHAPER W. 0.4"

ROSE CHAFER This slim, hairy beetle is one of the many scarab beetles, a large family of over 30,000 species, diverse in size and appearance. Some are important pests. The rose chafer feeds on roses, grapes, and a variety of other plants. Adults appear in late spring and early summer, eating both leaves and flowers. The larvae burrow into the ground, where they feed on roots, especially those of grasses. Control of the rose chafer is difficult.

JAPANESE BEETLE W. 0.4"

JAPANESE BEETLE It has been reported that when these Japanese, insects were first discovered on plants in New Jersey in 1916, experts could scarcely find a dozen. Now thousands can be collected daily, and control is a serious problem. The small, white grubs feed on the roots of grasses, damaging lawns. The larvae dig deep for winter and pupate the following spring. The adults emerge in midsummer and feed on cultivated plants and fruits. After mating, eggs are deposited in soil.

AMERICAN BURYING BEETLES W. 1.2"

CARRION BEETLE W. 0.7"   BLACK CARRION BEETLE W. 0.5"   HAIRY BURYING BEETLE W. 0.7"   HAIRY ROVE BEETLE W. 0.8"

CARRION AND ROVE BEETLES

CARRION AND ROVE BEETLES These beetles form a family of over 100 species, in two groups; the carrion and the burying beetles. The former are smaller, flattened insects which, as scavengers, feed on decaying animal matter. Some kinds are predators, feeding on worms and insects; a few eat plants. Both larvae and adults have similar feeding habits. The eggs are deposited on the corpse. The larvae of some species develop rapidly, reaching maturity in about a week. Rove beetles, which are also scavengers or predators, have short wing covers and superficially resemble the earwigs. Some live in fungi or in ant's nests. Some squirt a malodorous mist or droplet at enemies.

CLICK BEETLES The eyed elater is a striking example. If it falls or lands on its back, it lies quietly for perhaps a minute. Then, with a loud click, it flips into the air. If it is lucky, it lands on its feet and runs away; otherwise it tries again. The larvae of click beetles, known as wireworms, live in the ground or in rotten wood. Most click beetle larvae feed on roots, injuring potatoes and other crops. Some eat other insects.

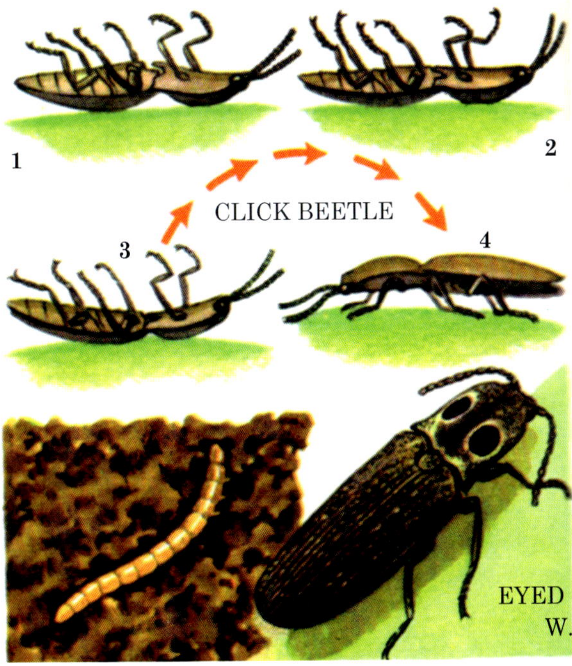

EYED ELATER
W. 1.4"

CATERPILLAR HUTER W. 1.2"

CATERPILLAR HUNTERS These are a group of fairly large and very attractive beetles. The family to which they belong is also very large and is related to the tiger beetles. These beetles are all predaceous, feeding on insects and other small animals. Larvae of the caterpillar hunters attack and feed on caterpillars of the gypsy moth and tent caterpillars.

FIREFLIES These are not flies at all, but soft bodied beetles, and most unusual insects. Widely distributed in Asia, America and other countries. The light giving property, or luminescence, is not confined to the adults. In some species the eggs and larvae glow also. The females of some species are wingless; these are known as glow worms. Fireflies are of little economic importance. They add to the pleasure of a summer's night the tropical display of thousands of these insects flashing in unison is breathtaking. Fireflies have posed a problem which scientists have not yet solved that of ``cold" light. The study of this phenomenon may have wide practical applications. In the species shown, it is the last segment of the abdomen which contains the light producing

Larva

Female

FIREFLIES W. 0.5"

tissues. This is very fatty and includes a network of nerves and airtubes. Through the latter, oxygen for the light producing process is obtained. In ordinary rapid oxidation much more heat than light is produced, here, heat production is negligible.

The larvae live underground or in rotted wood or rubbish. They feed on small insects. The adults are reported to eat the same food, but some of the common fireflies may not feed at all in the adult stage.

LADYBIRD BEETLES These are probably the best known and most valued of our beetles. The family is world wide in distribution. It has been remarked that the name can be traced back to the Middle Ages, when these beetles were dedicated to the Virgin - hence the name ladybird or ladybug. Both larvae and adults of many species feed on aphids. About 3,000 beetles will protect an acre of trees. The common species are generally similar in appearance but differ in the number of spots. All have very short legs, which distinguish these beetles from other garden beetles. Ladybird beetles lay their eggs on plants infected with aphids or scales. The larvae feed on the aphids and pass through four growth stages. When mature, they pupate in the remains of the last larval skin. Adults assemble by the thousands before cold weather sets in and hibernate under fallen branches or rocks.

LADYBIRD BEETLES

GARDEN BEETLES

Garden beetles are pests of the garden as well as the farm. They are the ones we spray, dust and pick. Other beetles are destructive in the garden also, as the May beetle, Japanese, beetle, and several weevils. These beetles illustrate the point that any cultivated crop is a bonquet for the right plant eating insects.

EPILACHA OR HADDA BEETLES Two species of Epilacha or Hadda beetles viz *Epilachna viginitiocto punctata* (28) and *Epilacha duodeca stigma* (12) are common. *E.* 28 *punctata* are deep red and have 14 black spots on each elytron whose tip is somewhat pointed. E. 12 *stigma* are deep copper coloured have 6-black spots on each elytron whose tip is more rounded. These attack solanaceous plants like brinzal, tomats and potato Damage is caused by beetles as well as grubs. Yellowish grubs have six rows of long branched spines.

MEXICAN BEAN BEETLE It is related to the common ladybird, and is one of the fairly large group of the ladybird family that feed on plants. Eggs are laid on the undersides of leaves. Spiny, yellow larvae eat the soft leaf tissue, leaving the veins behind. They eat pods too, stripping a plant in short order. Adults have similar feeding habits. Bean beetles feed on members of the pea family, wild and cultivated - peas, beans, alfalfa, and soybeans.

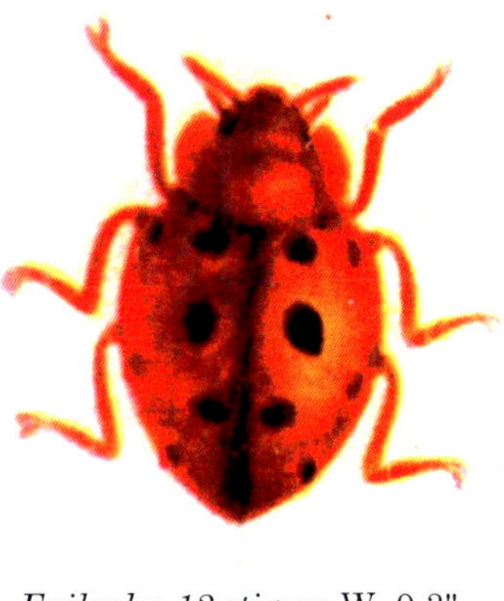

Epilacha 12 stigma W. 0.3"

E. 28 punctata W. 0.3"

MEXICAN BEAN BEETLES W. 0.3"
Eggs and Larva

STRIPED BLISTER
BEETLE W. 0.7"

TWELVE-SPOTTED
CUCUMBER
BEETLE W. 0.3"
STRIPED
CUCUMBER
BEETLE
W. 0.3"
Larva in roots of corn
Larva

CUCUMBER BEETLES The striped and spotted cucumber beetles are common garden pests. The larvae of the former attack roots, and the adults eat leaves of cucumber, squash, and related plants. The 12 spotted cucumber beetle is an even worse pest, feeding on many other plants besides those of the squash family. It appears early in the season and stays late. The larvae attack the roots of corn, oats and other grasses.

ASPARAGUS BEETLES The two species of them, ravage the asparagus crop. Adults hibernate in the ground, emerging in spring to feed on young shoots. Eggs are soon laid and larvae attack shoots, ``leaves" and fruit, stripping the plant. The life cycle takes about a month, so there are several broods a year.

CONTROL OF GARDEN PESTS Control of garden insects first requires a knowledge of which pests are involved and something of their eating habits. If you cannot identify the insect, seek aid of your country agent, college of agriculture. Once you know the insect, proper control measures can be learned from the same sources.

WATER BEETLES The whriligig beetles, true to their name, whirl or swim at the surface. They dive when disturbed and are good fliers also. Eggs are laid on water plants; larvae feed on water animals. Eggs of diving beetles are deposited in the tissues of water plants. They hatch into larvae commonly known as water tigers, which attack water insects, small fish, and even one another. After a month or so, larvae pupate in the ground. Adults are active all year. Their method of carrying a bubble of air down with them at the tip of their abdomen is interesting to watch.

The water scavengers, the largest water beetles, are sometimes over 3 inches long. Eggs are laid in a silken wrapping, attached to a floating leaf. The predaceous larvae feed like those of diving beetles. Pupae form in soil by late summer, and emerge in about 2 weeks. Adults also are active in all seasons. They carry air as a film on the underside of the body.

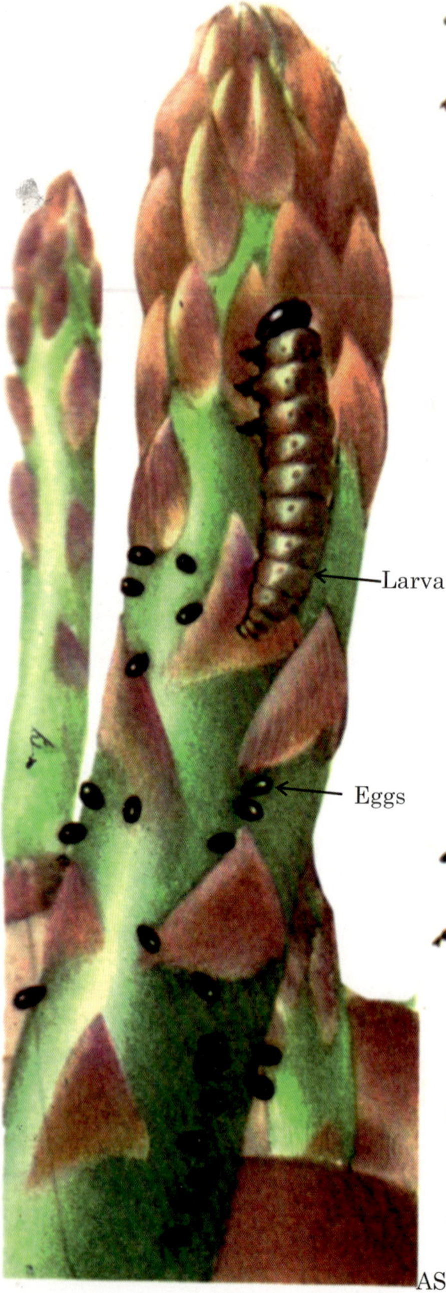

ASPARAGUS
BEETLE W. 0.3"

SPOTTED
ASPARAGUS BEETLE W. 0.3"

DIVING BEETLE W. 1.1"
MAYFLY Larva
DIVING BEETLE larva
WHIRLIGIG BEETLES W. 0.6"
WATER SCAVENGER W. 1.3"

MAY BEETLE W. 0.9"

LARVAE

MAY BEETLES These form a group of over 100 American species, widely distributed and difficult to control. White eggs are laid in an earth covered ball amid roots on which the white grubs feed for 2 or 3 years. These pupate underground in autumn and adults appear the following spring. Adults feed on leaves of many common trees.

GREEN JUNE BEETLE This beetle feeds on many plants, eating roots, stems, and leaves. Larvae are found in soil or manure. They move by bristles on their backs instead of by their short legs. Adults fly in large numbers, making a loud buzzing which is somewhat similar to the buzzing of bumblebees. The adults damage apricots, figs, grapes, melons and other fleshy fruits.

GREEN JUNE BEETLE
W. 0.9" and larva

SCARAB BEETLES These beetles including the rose chafer, Japanese beetle, and may beetle, form a large family totalling more than 30,000 species. Many are scavengers, adapted for living in or on the ground. Larvae are usually large white grubs found in the soil. Of the many scarabs, the dung beetles or tumblebugs are outstanding. The adults form balls of dung and roll them about giving the impression of being industrious workers. Eggs are laid in the ball, which is buried. The ferocious looking rhinoceros beetles and their relatives, the ox beetles, are the largest of the scarabs. All are harmless. Some are 2 inches long; much larger tropical forms occur. Males have more prominent horns than females. The larvae are found in rotted wood of rich soil.

DUNG BEETLES
W. 0.8"
Female
Male
UNICORN BEETLE
Female
TUMBLE BUGS W. 0.8"
OX BEETLE W. 1.0"
HEMICCEROS BEETLE
W. 1.0"
UNICORN BEETLE
W. 1.5" Male

HORN BEETLE (1.4"w)

Head of female

STAG BEETLE male (1.6" w)

STAG AND HORN BEETLES These are related to scarabs. The large stage are so named because the huge mandibles of the males resemble the antlers of stags. Mouth parts of the females are much smaller. Large white grubs are found in rotted wood. Horned beetle larvae and adults are often found in large colonies in burrows in rotted logs. These beetles make noise by rubbing wing-covers or legs. The adults are harmless.

INDIAN RHINOCEROS BEETLE

The adult beetle is known as ``Rhinoceros beetle" because it has a pointed horn on the head. The body of the beetle is robust and elongate. Unlike other beetles, the wings are well developed and it is a good flier. The adult beetle is one of the most destructive pests which feeds on the crown of the coconut trees. The larvae are harmless.

Side view of male

Dorsal view of male

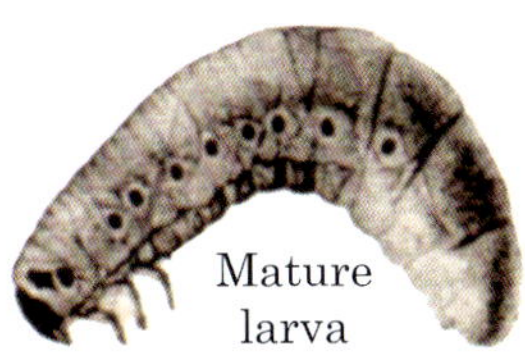

Mature larva

INDIAN RHINOCEROS BEETLE
W. 2"

RED PUMPKIN BEETLE

RED PUMPKIN BEETLE The adult is a brilliant orange coloured. beetle, with the body surface shiny and smooth. The underside is entirely black and is covered over with short, wide, soft hair like setae. It is a serious pest of cucurbitaceous plants like melon, pumpkin, cucumber etc. The attack caused by the feeding of adult insect or its larva on leaves, is indicated by the formation of irregular holes in the leaves. The larvae also damage the shoots, roots of the plants which ultimately withers and dies. The chief damage however is done to the fruits which are rendered useless for human consumption.

CIGARETTE BEETLE

It is commonly known as the cigarette beetle and is a polyphagous pest which infests a vast variety of commodities for example turmeric, ginger, opium, tobacco, saffron, redpepper, cumin seed, nux vomica seed, belladonna etc. This pest is distributed all over the world. The adult beetle is small and oval, light to dark brown in colour. It is active during afternoon and night. It is phototactic. The longevity of the adult beetle is 4 or 5 weeks or more depending on climatic conditions.

CIGARETTE BEETLE
(0.2-0.3" W)

BANDED / STRIPED BLISTER BEETLE

BANDED / STRIPED BLISTER BEETLE W. 12"

The banded or striped blister beetles of meloidae are provided with red and black bands, six in number. The meloid beetles yield cantharidin, a crystalline solid from dried beetles, used for preparation of several medicins including hair oils. The species are common on the red and yellow flowers but are not the serious pests.

DESERT BLISTER BEETLE (1.5" W)

DESERT BLISTER BEETLE

The desert blister beetles are found only in deserts. They are fairly long. The head and prothorax are red in colour where as the elytra are black with white spots. When disterbed, the blister beetles exude an acidic fluid. The exudation may cause blisters on the skin.

METALLIC WOOD BORERS

Thrincopyae ambiens W. 1.5-2"

These belong to family Buprestidae and are commonly called metallic beetles. Minute or large in size and flat or cylindrical in shape. This group have two significant features which make them noticeable, one is that their larvae turnnel in the wood of living trees, sometimes killing the trees and often spoil timber. On the other hand some species are extremely beautiful metallic coloured often golden, bronze, green or blue. The adult are lovers of bright sunshine and inhabit wooded areas. The beautiful and iridescent coloured elytra have attracted the attention of man from very

Buprestris aurinta W. 1.5"

early times and the brilliant metallic Buprestide beetles have been utilized in ornaments. Embrydery and other work of art in India, China, Japan, Africa and other countries e.g. *Sternocera chrysicioiolis, S. acqusignate, S. punctata foveata* etc. are much sought for jewellers. Beautiful garlands are made by using various types of metallic beetles.

Buprestris sp. W. 2"

Strenocera chrysoli W. 1.5"

Buprestris sp. W. 1.5-2"

S. acquisignata W. 1.5"
S. punctato foveata W. 1.5"

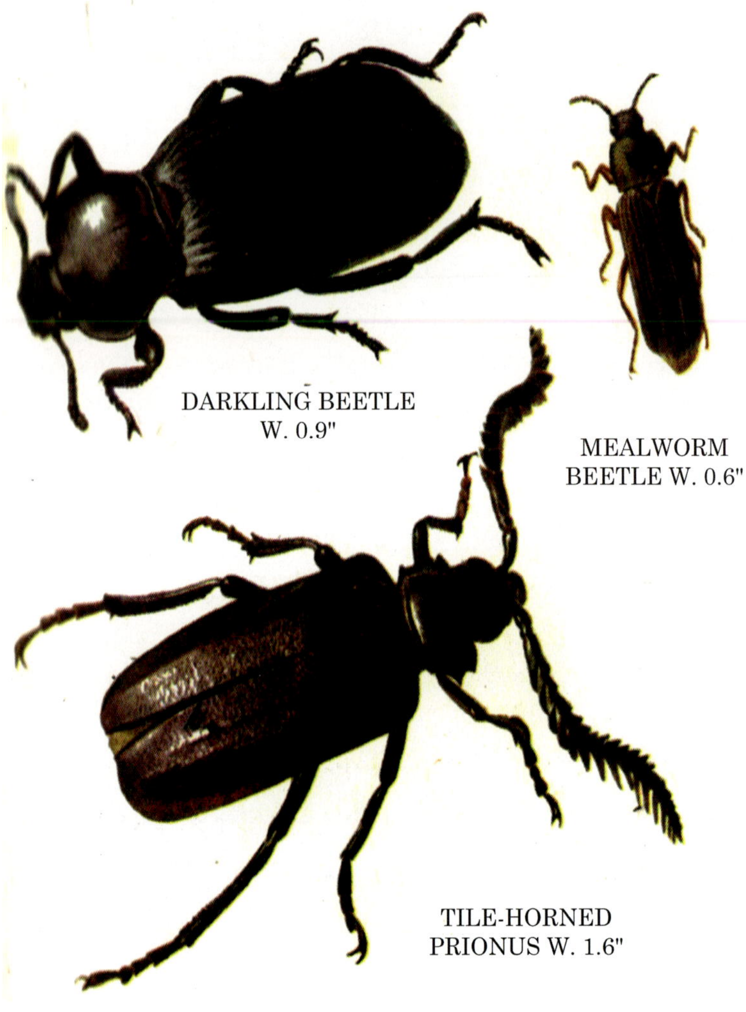

DARKLING AND PRIONUS BEETLES Darking beetle larvae are the ``mealworms" feed on stored grain and hence are serious pests. There are a large number of closely related species. The larvae of prionus beetles are known as round-headed borers. These attack roots of fruit and ornamental trees, grape, and other plants. The antennae, with overlapping plates, slightly longer in the male, identify the prionus beetles.

FLAT-HEADED BORER W. 0.8"

FLAT-HEADED AND LONGHORNED BORERS

Larvae of many beetles bore into wood but flat headed borers (Bupretids), and longhorned borers (Cerambycids), represented by the locust and elder borers and the pine sawyer, are vicious pests of orchard, shade and forest trees. Larvae of flat headed borers feed mostly just beneath the bark, whereas longhorned borers usually channel far into the trunk. The attractive locust borer adult is often found on goldenrod.

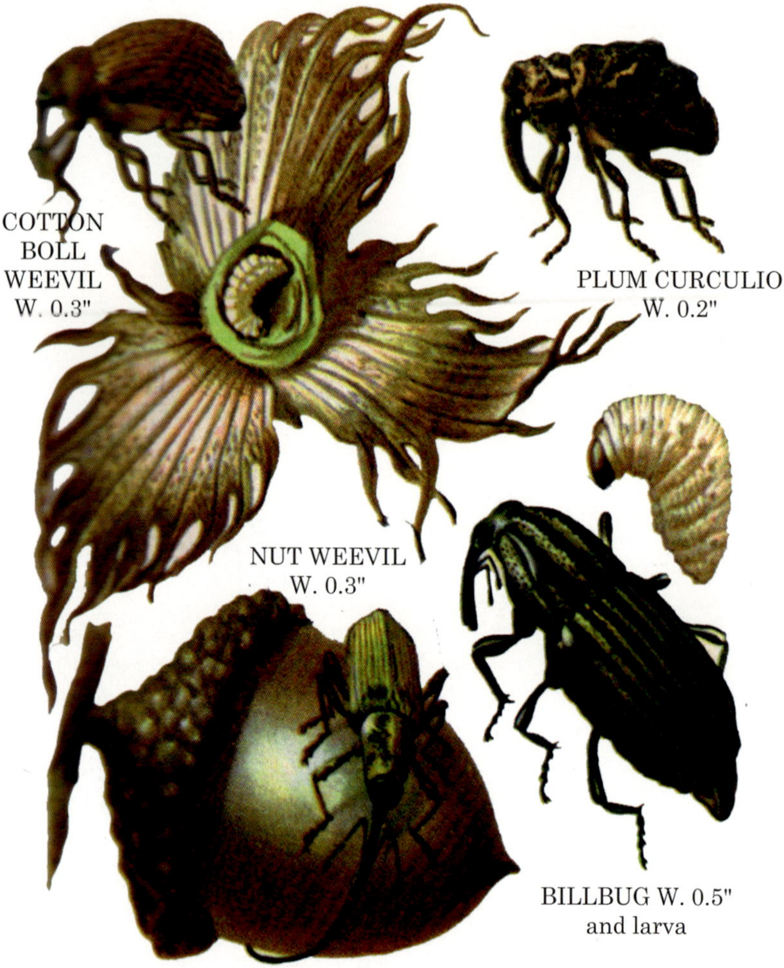

WEEVILS These are small beetles with mouth parts modified into a downward curving beak or snout. Many are important pests of grain. The plum curculio damages peach, cherry and plum trees. Nut weevils are found in acorns and all other edible nuts. The granary weevils, found in grain, are prolific with a life cycle of only 4 weeks. Some feed on the roots, others on stored grain.

HORNTAILS AND SAWFLIES These closely related, belong to the bees and wasps but lack the constricted abdomen. Horntails lay their eggs on dead or dying trees. The larvae are borers, which pupate in their deep tunnels. Most sawfly larvae feed on leaves. Horntails are hosts to the ichneumon flies, which parasitize the larvae. The female ichneumon fly can locate a horntail burrow under several inches of wood and deposit her eggs therein.

PIGEON HORNTAIL W. 1.9"

ICHNEUMON FLIES These flies play an important role in controlling many harmful insects. They are more closely related to wasps than to flies. Their larvae are parasites of caterpillars and of larvae of beetles, flies, and other pests. The female ichneumon fly, with her unusually long ovipositor, attracts attention and is sometimes feared by those who do not know she is harmless. This ovipositor can pierce several inches of wood.

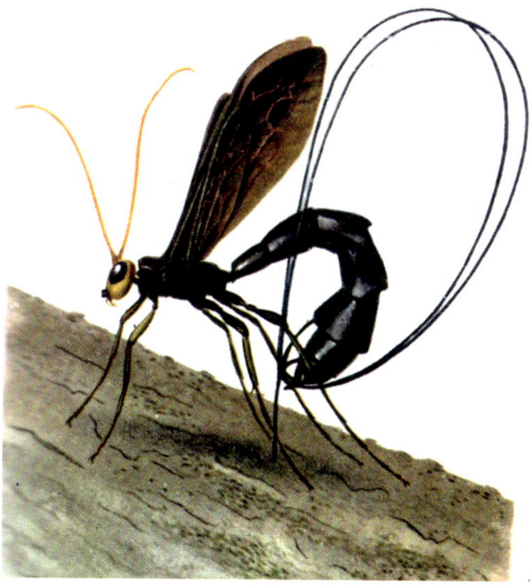

ICHNEUMON FLIES W. 1.4"

CARPENTER ANTS W. 0.5" workers

CARPENTER ANTS: All carpenter are social insects, living and working together in ways that have astonished laymen and naturalists alike. Carpenter ants and their relatives form one of the largest groups of ants. They build nests and burrows in dead wood, logs, and the timbers of buildings, where they may do considerable damage if allowed to spread. Carpenter ants are world wide in distribution. The workers, which are infertile females, are among the largest known ants.

FIRE ANT W. 0.2"

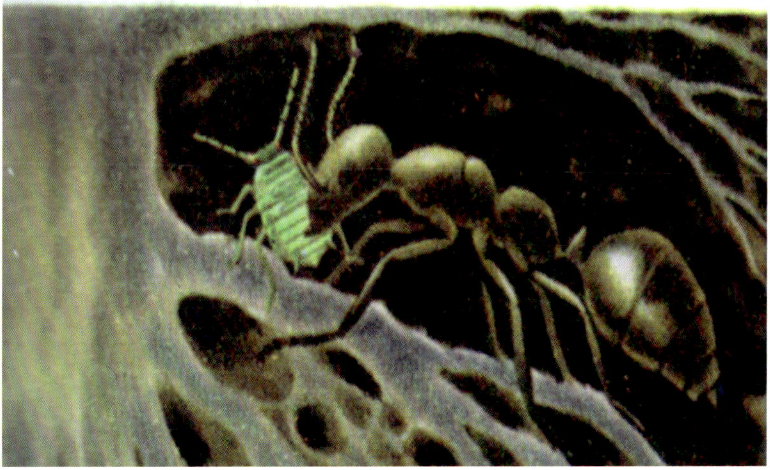

CORNFIELD ANT W. 0.2"

FIRE AND CORNFIELD ANTS These ants eat the sweet secretions of corn root aphids. Aphids lay eggs in the ant burrows. When these hatch in spring, the ants place the aphids on knot weed roots till the corn is planted and growing. Then the ants transfer the aphids to the corn roots, thus insuring a constant food supply. Cornfield ants are widely distributed and very abundant. It has been reported that they are the most abundant of all our insects, a fact which makes these small ants important.

HOUSEHOLD ANTS Pharooh's ants are small but numerous. They are common invaders of homes, feeding on any sweet footdstuffs. The Argentine ant, is a serious household and garden pest. They may invade nests of other ants and hives of bees. Fortunately they are semitropical and are limited in their movement northward. These ants protect aphids to secure the honey dew they produce. Little black ants are found outdoors more than within.

INSECT GALLS

INSECT GALLS Small, wasp like insects (Cynipids), flies, and others lay eggs in plant tissues. Each insect selects a specific plant. As eggs hatch, the plant tissues around the larvae begin to swell, forming a characteristic gall. The larvae feed on plant juices and pupate in the gall. The adult emerges by burrowing through the side. Some galls are large and woody, some soft, some `knobby' and spiny. Best known galls are the oak apples and the galls commonly seen on roses, black berry, and goldenrods.

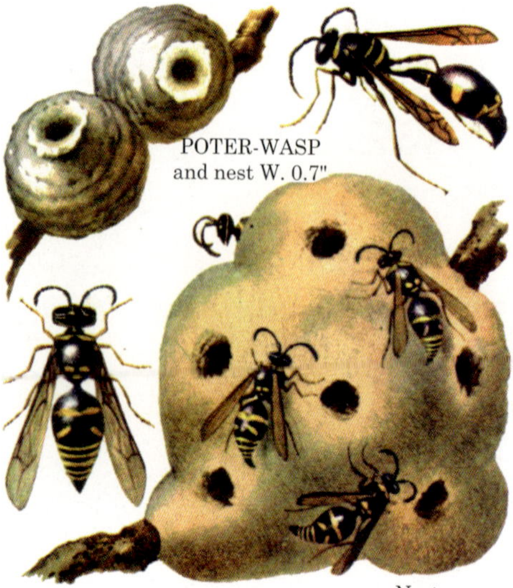

MASON WASP W. 0.5" Nest

MUD WASPS Several families of wasps are represented among those building their nests with mud. Potter wasps are solitary, each building a vase shaped nest of mud on plants. These wasps prey on caterpillars and beetle larvae. Most mason wasps, in the same family as the potters, nest in burrows in the soil, but the species illustrated here makes a clay nest on a branch. Best

MUD DAUBER W. 1.0" and nest

BLUE AND DAUBER W. 0.8"

known mud wasps are the mud daubers, which make large nests on walls in attics or deserted buildings. The female builds the nest of many mud cells. In each she places several paralyzed spiders or other insects before she lays the egg and seals the cell. The blue mud dauber uses nests made by the common mud dauber. The female moistens the cell wall, digs through, removes the contents, and refills the cell with her own spiders and egg. The cuckoo wasp, named after the European cuckoo, awaits its opportunity and lays its eggs in the nest of a mud dauber while the latter is off searching for a victim. The cuckoo wasp larvae feed on insects provided for the young mud dauber.

VELVET ANTS Velvet ants are so named because the wingless females are ant like. They are parasites of wasps and bees, specially of the solitary species. The female wasp can really sting, but the males do not. The female crawl down the burrow and kills the owner with a powerful sting. She then lays her egg on the owner's larva, which her own larvae later eat. The solitary, thread-waist wasp illustrated below is one of several kinds parasitized by velvet ants.

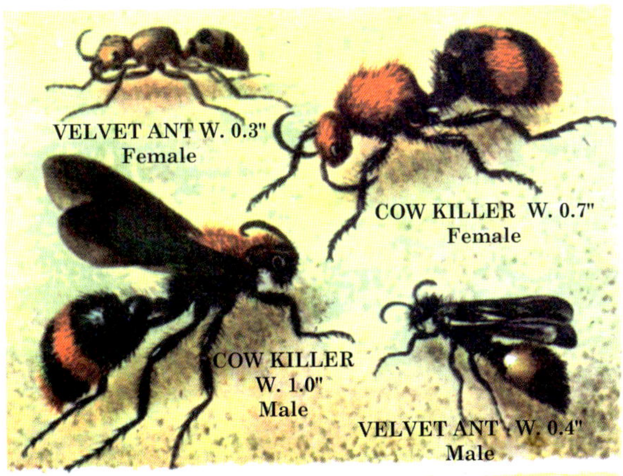

EUROPEAN HONEY BEE W. 0.6"

HONEY BEES

There are several kinds of social bees which store honey. The most important are the stingless bees and the honeybees. The stingless bees are tropical occurring in Central and South America, Africa, Asia and Australia. These bees nest in hollow trees, caves or other cavities and store considerable honey which is sought mainly by man and monkeys. In India four forms of bees are commonly found viz. *Apis mellifera,* the famous European honeybee; *Apis dorsata* the giant bee of India, *Apis florea* the dwarf bee and the Indian honey bee *Apiscerana indica.*

Honey bees are trimorphic having Queens (fertile female) drones (males) and workers (un fertile females) in their colonies which perform. Their duties, Further, it is interesting to note that it requires about 40,000 to 80,000 trips of a honeybee, and visits to many times this number of flowers, to find and assemble nectar enough to make about 500 gm. of honey. Each trip is about 2 or 3 km. Thus for a single bee to collect nectar enough for 1/2 kg of honey would mean travelling at least double the distance around the world.

All the bees viz., Bumble bees, Sweet bees, leaf cutting bees and honey bees are distinguished from wasps in a number of minor ways. Their legs have ``pollen baskets" of stiff hairs, and the body

INDIAN GIANT HONEYBEE W. 0.8"

is hairy also. Bumblebees are larger than most others. Their elongated mouth parts enable them to pollinate red clover, which no other bee can do. Nests are made underground by the fertile females (queens), which survive the winter. The colony consists of a queen, workers and drones. Sweat bees, small and brilliantly coloured, nest in the ground. They are attracted to perspiration; hence their name. Leaf cutting bees are larger and brightly coloured also. Their nests, made underground, are lined and divided with leaves which the bees have cut in ovals or circles from roses and other plants. The honeybees are probably the best known of all the insects. Honey has been obtained from wild and from kept colonies far back into history. In making honey, bees pollinate However, fruit trees and other plants. A normal colony of a queen, workers, and drones may contain up to 50,000 bees. Periodically bees swarm, and the old queen goes off to found a new colony, leaving a young queen behind.

INDIAN DWARF HONEY BEE
W. 0.3"

INDIAN HONEY BEE W. 0.5" →

Apis cerana indica W. 0.5"

CARPENTER BEE W. 1-2"

CARPENTER BEE : The carpenter bee is usually larger, densely hairy, black, dark metallic blue. It is cosmopolitan. These bees are without pollen gathering apparatus. These bees often cause great damage to wooden rafters but also does considerable good by pollinating flowers.

MUSTARD SAWFLY Adult has yellow, brown markings on the body and dark reddish brown colours on the wings. Damage is done by larvae which are harmful pest of mustard, cabbage, cauliflower, turnip radish etc.

MUSTARD SAWFLY W. 1.2"

CICADA KILLER W. 1.6"

BURROW OF CICADA KILLER

CICADA KILLER This large solitary wasp digs a burrow a foot or so deep. In side passages the female stores adult cicadas which she has paralyzed by stinging. The heavy cicadas are dragged up a tree by the killer till she can get enough altitude to fly back to her burrow. When the egg hatches, the larva feeds on the helpless cicada. In a week it is full grown and pupates in a loose cocoon. It emerges the following summer, completing its life cycle.

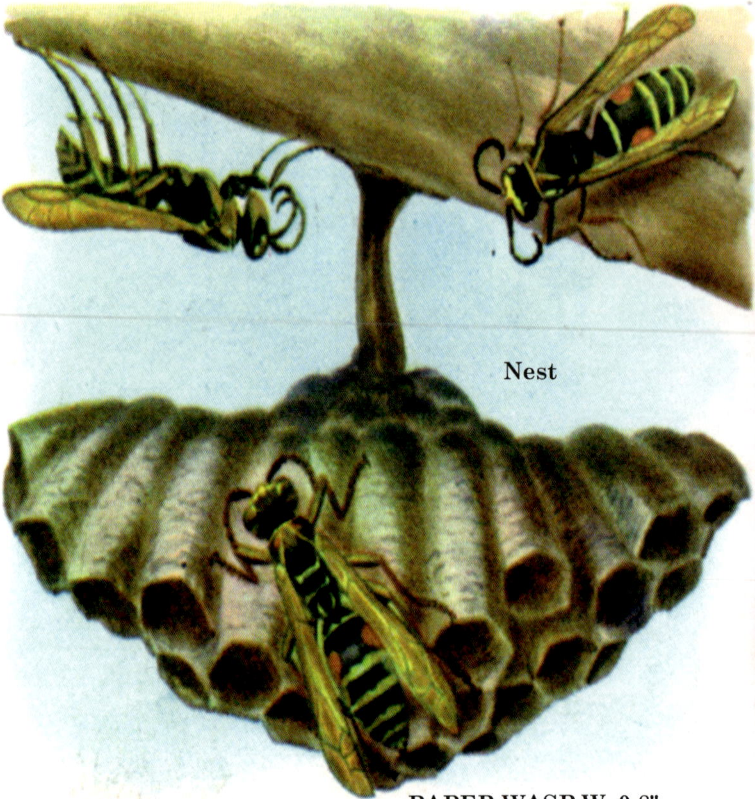

PAPER WASP W. 0.8"

PAPER WASPS: These are the common wasps everyone learns to know sooner or later by painful experience. The common paper wasp builds on unprotected paper nest out of wood it chews up. The nest is hung under eaves or in barns, or in other sheltered sites. No food is placed in these nests. After the eggs hatch, the young are fed daily till they pupate. The bald faced or white faced hornet builds an oval, covered nest; some nests can accommodate over 10,000 hornets. Unfertilized eggs develop into drones or males. Fertilized eggs grow into workers or queens, depending on the food eaten. Queens make small brood nests in early spring. Yellow jackets, which are closely related build nests of varying size, some underground, some hidden in rock walls or under logs. An empty fieldmouse nest is often used.

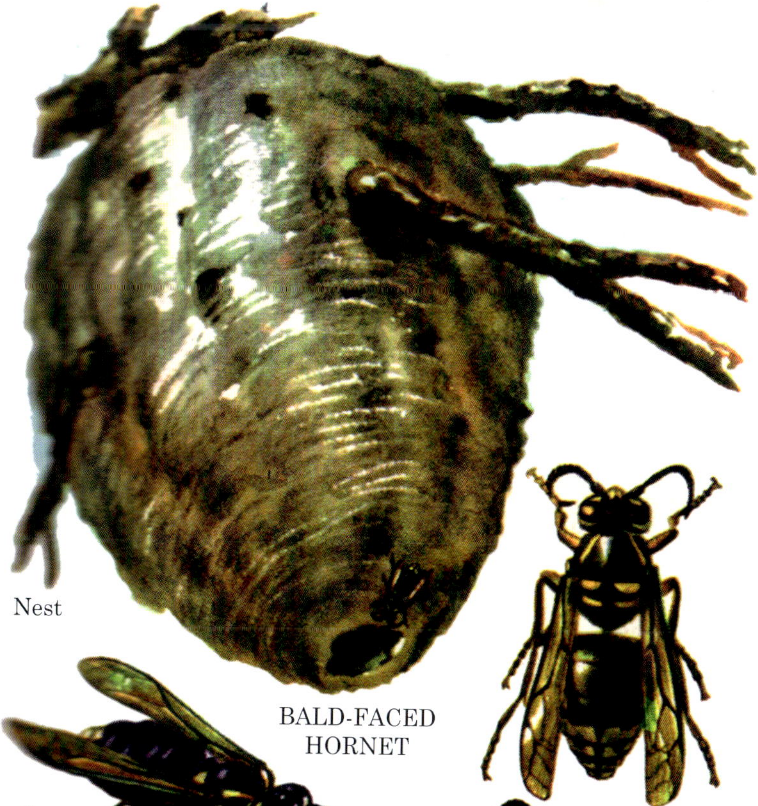

Nest

BALD-FACED HORNET

Drone
W. 0.7"

Worker
W. 0.6"

Queen W. 1.0"

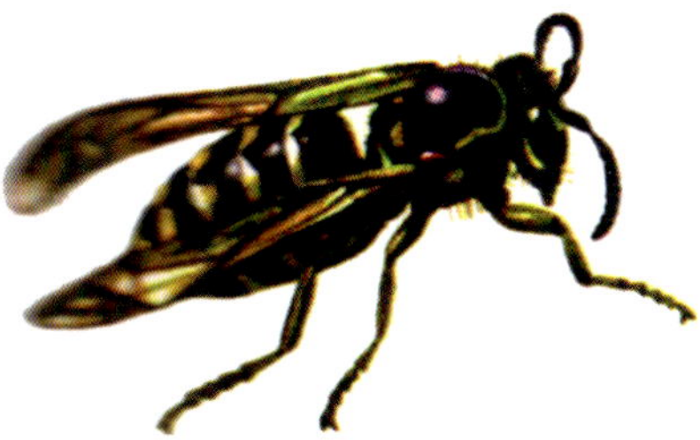

YELLOW JACKET W. 0.8"

PESTS OF CLOTH AND CLOTHING: The carpet beetles are closely related to lardder beetles and to other species which destroy specimens in museums. They eat all kinds of animal matter, thriving on rugs, woolens, fur, leather, and hair. The larvae which do the damage can be controlled by chemical repellents, poisons, and sprays. Adult clothes moths do no harm. Females lay white oval eggs on clothing, leather, etc. These hatch in about a week into larvae, which may feed for a year on your prized possessions before they pupate.

PESTS OF ANIMALS: There are many and cause severe losses. Dog fleas spread from dogs to people and may, in some areas, carry plague. They also infest other domestic animals. Control by dusting. Other species are similar. The sheep tick, a species of fly, feeds by sucking blood. The eggs hatch in the body of the female, and the larvae develop to maturity before being deposited. Control is achieved by shearing and dipping the sheep. The squalid duck louse is typical of bird lice which live on wild and domestic birds.

CARPET BEETLE
W. 0.2"
BLACK CARPET BEETLE
W. 0.3" and larva
CLOTHES MOTH W. 0.5"
Larva
DOG FLEA W. 0.1"
SLEEP
STICK W. 0.2"
DUCK
LOUSE
W. 0.2"

OTHER COMMON PESTS

Insects that do us harm have been freely scattered through the pages of this book. Some have said that these six legged creatures are man's worst enemies. Here are more we should know to be forewarned and so forearmed.

SILVER FISH W. 0.4"

SILVERFISH The most primitive insect, yet still capable of eating starch from book bindings, wallpaper, and clothing. Common indoors, especially in warm places.

BED BUG W. 0.3"

BEDBUG This flat bodied household pest, once established, may spread rapidly. May have 3 or 4 generations annually. Female lays white eggs in cracks; eggs hatch in a week or so. Destroy with Lindan or other spray.

HOUSE FLY This is the most common and most despised of all the flies. Public health campaigns have not destroyed it. Associated with garbage and filth, flies can reproduce a generation a month when temperatures are favourable.

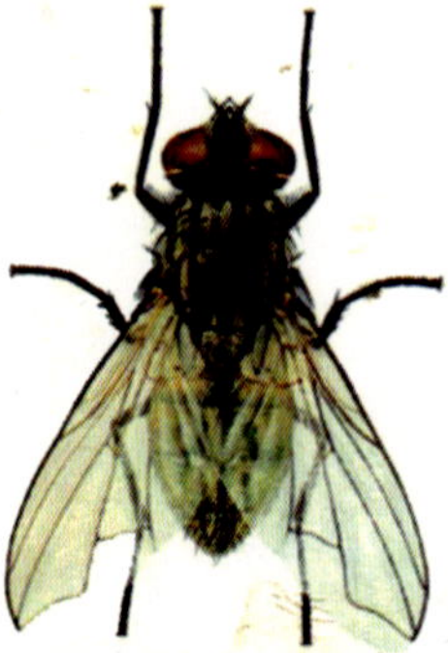
HOUSEFLY W. 0.5"

BOOK LOUSE W. 2"

BOOK LOUSE Found in Collections of old books and papers where they remain hidden and feed upon paste and paper. These are tiny soft bodied wingless light coloured insects with chewing mouth parts and long legs and antennae.

THRIPS

Thrips are blown by wind into the green houses, nurseries and elsewhere and in this way they constitutes a complex of species that are involved in the injury caused to the floricultural crops.

Damage is caused by the thrips in several ways. They may directly injure the flowers with their rasping sucking mouth parts. They vary in colour from brown to yellow. Their wings are narrow provided with fringe around the margins. Immature forms are lighter in colour, smaller in size than the adults and without wings.

HOUSE CRICKET These insects are as common as the cockroaches and are also nocturnal in habit. They make a very shrill noise by their stridulatory apparatus. They live in various crevices in the walls, floor, ceillings, doors, furniture, and books etc. For control use 5% dust of malathion.

FLOWER THRISP W. 0.2"

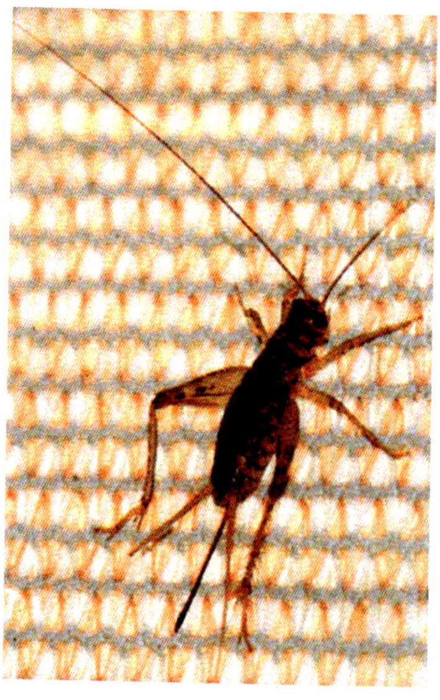

HOUSE CRICKET W. 1.4"

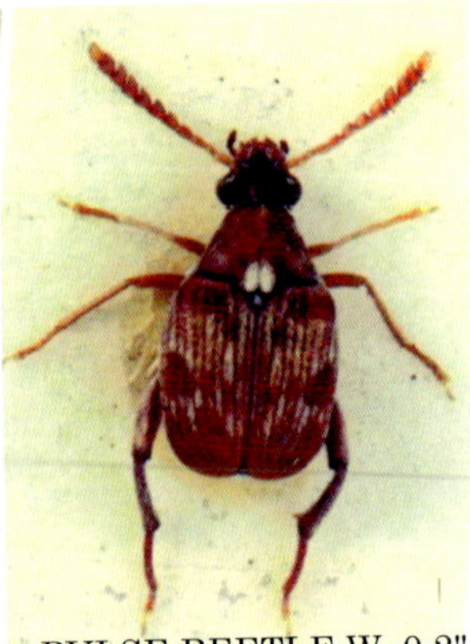

PULSE BEETLE W. 0.2"

PULSE BEETLE : It is a serious pest of stored pulses. All kinds of pulses are eaten by this pest and stored pulses may be completely destroyed.

All the pest of the stored grains can be controled by fumigation with carbon disulphid, mathyl bromide chloropicrine, aluminium phosphide etc. The other graine protections are malathion 1%, piperonyl butoxide or repellants or antifeedants etc.

LESSER GRAIN BORER W. 0.2"

LESSER GRAIN BORER: The lesser grain borer does almost as much damage as the rice weevil. It feeds in both larval and adult stages in the interior of nearly all types of grains. It can also ingest flour. It is most common in wheat and is one of the most destructive wheat insects.

KHAPRA BEETLE 0.1"

KHAPRA BEETLE: This is perhaps the hardiest among the insects that infest stored grains. This beetle feeds on most grains and sometimes on pulses. Wheat flour is its best food and only the larvae of this insect are destructive while the adults are harmless.

RICE WEEVIL 0.1"

RICE WEEVIL : It is one of the most destructive stored grain pests. These insects frequently cause complete destruction of grains and the infested grains will usually be found to be heating at the surface, and it may be damp, sometimes to such an extent, that sprouting occurs. Normally the kernels of the grain are eaten. Adults feed upon whole seeds or flour but the larvae develop only in seeds or in pieces of seeds large enough for them to live within.

RUST RED FLOUR BEETLE : It feeds mostly on broken grains or grains damaged by other pests. It is also a pest of prepared cereal products like Atta, Maida and Suji. A heavy infestation of products by these insects gives a yellow colour to the food which becomes mouldy with the excreta of larvae and adults. Both larvae and adults attack the grains.

RUST RED FLOUR BEETLE 0.2"

REFERENCES

Anonymous (2008). Amaging book of facts. Paragon Books, U.K.

Blanely, W.M. (1977). How. The insects live. Galley Press.

Corti, W.R. (1964). Butterflies and moths, The Odyssey Library, New York.

Fletcher, T.B. (1914). Some South Indian Insects. Govt. Press, Madras.

Gooddon, R. (1971). Butteflies. Hamlyin, London.

Gooddon, R. (1973). All colour book of Butterflies. Octopus Books.

Goodden, R. (1977). The wonderful world of butterflies and moths. Hamlyin, London.

Gulan and Cranton (1965).

Lefroy, H.M. (1909). Indian insect life. Thecker Spink and Company, Calcutta.

Mani, M.S. (1968). General Entomology. Oxford & TBH Publishers Co. New Delhi.

Metcalf and Flint (1962). Destructive and useful insects: Their habits and control. M.C. Graw Hill Book Company, New York.

Pradhan, S. (1969). Insect pests of Crops. National Book Trust, India.

Russel, A.S.H. (2007). Whitakers World of facts. Penguin Books.

Tweedie, M. (1973). All colour book of insects. Octopus Books.

Zim, H.S. and C. Cottom Ed. (1956). Insects: A book to familiar American insects: Golden Press, New York.

Appendix - 1
AMAZING INSECTS

Most abundant: Insects called springtails live in topsoil all over the world. There are probably as many as 600 million per hectare. Together they weigh more than the entire human race.

There can be up to five billion aphids per hectare in a swarm.

A swarm of desert locusts may contain 50 billion insects. One seen in 1889 was reckoned to have 250 billion locusts weighing a total of half a million tonnes.

Fastest: Dragonflies can fly at speeds of 28.7 - 32 km./h.

Heaviest: Goliath beetles can be 11 cm long and weigh 100 g.

Longest Wingspan: The giant silkmoth *Attacus atlas* is found from India eastward through China to Malaysia. It is reputed to be the largest moth in the world. It attains a wing expanse of 12 inches. Likewise, the Queen Alexandra's birdwing of Papua New Guinea is the largest butterfly which can be 12 inches across.

Longest: The world's longest insect is the giant stick insect of Indonesia which grows upto 13 inches long.

Smallest: The world's smallest insect is *Mymar* 0.2 mm long (a bettle of Trichopterigidae) parasitic in eggs of insects.

Longest-Lived: Jewel beetle larvae, or young, may live inside timber for 30 years or more before becoming adult insects.

Shortest-Lived: Mayflies may live for a single day as adults, but for two or three years as larvae.

Noisiest: Colonies of hundreds of male cicadas make loud noises to attract females. These sounds are as noisy as heavy traffic.

Long-distance fliers: Butterflies have been tracked travelling 4,828 km. (Goodden, 1977; Russell, 2007; Anonymous, 2008).

Appendix - 2

BEE FACTS

There are about 20,000 different types of bee. The best known is probably the honeybee, but not all bees live in colonies like the honeybee. Many live alone and build their own nests.

Wallace's giant bees are the world's largest at upto 4 cm long. They were first found in Indonesia in 1858 but were then thought to have become extinct. In 1981 the bees were rediscovered.

The smallest bee is the Brazillian *Trigona duckei* at 2-5 mm long.

Honeybees are the only insects that make food which humans eat. Bees make honey to feed the inhabitants of their hives during winter. The honey they do not eat is harvested by beekeepers. Bees also make beeswax which is used for making candles and furniture polish.

One beehive may contain 50,000 worker bees-these are the bees which collect nectar from flowers to make honey.

Honeybees visit upto five million flowers to make one kilo of honey. They fly a total distance equal to flying four times round the Earth.

Queen bees can lay two or three thousand eggs a day - as many as 200,000 eggs a year. They may live as long as five years, so can produce a million eggs.

Honeybees' wings beat 11,400 times a minute. These wing movements make the bees' familiar buzz.

1.3 million tonnes of honey are produced worldwide every year. China is the main producer with 276,000 tonnes, followed by the USA with 82,000.

Some people are allergic to bee stings, which can even be fatal. In 1962 in Rhodesia, Johanne Relleke was stung 2,243 times by wild bees, but survived (Russell, 2007).

Chemical composition of honey

Chemically, honey is a viscous water solution of sugars, containing 13 to 20% water, 40 to 50% fructose, 32 to 37% glucose, 2% sucrose, traces of maltose, 1 to 12% dextrins and gums; and traces of minerals, free organic acids, vitamin A and B-complexes, enzymes, plant-pigments, pollen grains and bee wax.

The value of the honey lies in its simple sugars found in large amount in it which are directly assimilable and are a ready source of bodily energy. The colour of honey and taste are no guides to the purity and quality of honey. Honey from soapnut and white clover, is light (almost water white) in colour and mild in flavour, that from mustard looks yellow and has a little pronounced flavour and that from shisham (*Dalbergia*) is dark amber and strong in flavour. Barberry and buckwheat honey are very rich dark in colour and have a strong taste.

Honey as food - It has been estimated that 200 gm of honey is as nourishing as 1.135 kg of milk or 1.658 kg of cream cheese or 340 kg of meat or 425 kg of boneless codfish and 8 oranges or 10 eggs.

It is interesting to note that it requires about 40,000 to 80,000 trips of a honeybee, and visits to many times this number of flowers, to find and assemble nectar enough to make about 500 gm. of honey. Each trip is about 2 or 3 km. Thus for a single bee to collect nectar enough for 1/2 kg of honey would mean travelling at least double the distance around the world.

A Manual of Practical Entomology (Field and Laboratory Guide)

3rd Revised Edition

M.M. Trigunayat

A Manual of Practical Entomology (Field and Laboratory guide) is written to provide text material on different aspects of the practical syllabi of M.Sc. Entomology. All chapters are illustrative and well explained. Easy text will help students understanding the exercises. Dissections of the individual insect have been described so as to understand and follow the finest details of the anatomy of the insect concerned. Physiological and Behavioural Exercises are well presented usually in the style followed by the students. Individual Orders have been dealt with relevant necessary text and vital aspects of biology of the insect concerned. Chapters on Natural History and Rearing will inculcate interest among the students for their quest on these tiny creatures. Profusely illustrated with simple figures presentation and style gives the reader an insight to make it unique. Primarily intended for use by the post-graduate and graduate students of Entomology of the Universities in Rajasthan and elsewhere in India, it can also be used by agriculture departments, naturalists and workers in other related fields.

Contents

ISBN: 9788172339838 | 2016 | 328 pages

Principles of Insect Behaviour

2nd Revised & Enlarged Edition

V.B. Awasthi

The aim of writing this book has been to provide an account of the principles of insect behaviour for the post graduate students of Entomology and general Zoology of all the Universities and Research Institutions as well as to the research scholars engaged in the study of different aspects of Insect behaviour. It is also useful to general readers who have basic knowledge of Biology and have curiosity to know why the insects are vividly coloured, have peculiar shapes, produce sound and light; how they communicate and why bees and wasps sting when disturbed. Significance of insects in pollination of plants associated with the production of seeds, gains and fruits, the role of hormones, pheromones and other chemicals controlling various vital events of the insect behaviour has also been described. The importance of insect migration, control of insect pests by behavioural means has also been discussed. Thus, this book deals with almost all major aspects of insect behaviour. It is hoped that this book will be useful not only to the students of Entomology-Zoology and research scholars of insect behaviour, but also to naturalists who wish to know about some of the mysteries of these small creatures and their lifestyle.

Contents

1. Kinds of Insect Behaviour; 2. Camouflage, Warning Colouration and Defensive Behaviour; 3. Social Behaviour and Parental Care; 4. The Language of Bees and their Behaviour; 5. Interactions between Insects and Plants as related to Pollination; 6. Food and Feeding Behaviour; 7. Intraspecific and Interspecific Interactions; 8. Adaptations in Insects; 9. Thermoregulation in Desert Insects; 10. Ecological, Behavioural and other Adaptations of Desert Insects; 11. Thermoregulation by Behavioural Means; 12. Photoperiodism and Insect Behaviour; 13. Photoperiodism and Ecological Adaptations; 14. Hormonal Control of Insect Behaviour; 15. Pheromonal Control of Insect Behaviour; 16. Genetic Control of Behaviour and Evolution of Behaviour; 17. Communication in Insects; 18. InsectPestManagement by Behavioural Means; 19. Migratory and Dispersal Behaviour of Some Insects; 20. Influence of Host Plants on Sex Pheromone Behaviour of Phytophagous Insects; 21. The Role of the Sense Organs in Mating Behaviour of Cockroaches; 22. Reproductive Behaviour; Selected Bibliography; Glossary of Selected Technical Terms; Appendix; Subject Index

ISBN: 9788172338251 | 2013 | 299 pages

Introduction to General and Applied Entomology *3rd Edition*

This text book aims at to provide an up to date, general and applied account of Entomology which is written in a simple and lucid language keeping in view the syllabi of Entomology in M.Sc. Zoology & M.Sc. (Ag.) of various Indian Universities. The book will also be useful to those who are preparing for competitive examinations conducted by various central & state Government agencies for recruitment. The book is equally useful to those dealing with pests and pest control. It is hoped the book will be useful not only to students but also to agriculturist, horticulturist, foresters, veterinary and clinical doctors and laymen dealing with pest control.

ISBN: 978-81-72335-97-7 | 2019 | 499 Pages

Insect Neuroendocrines

V.B. Awasthi

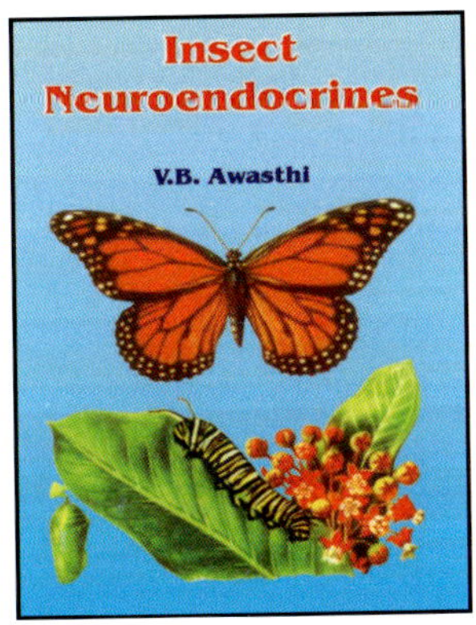

The aim of the author has been to present an account of the neuroendocrine system of insects, the super importance of which is well established, and also to provide information as far as possible directly, through the ultrastructural illustrations. Besides, an account of some of the hormonally controlled phenomena of vital significance is given.

It is hoped that the theoretical account, and a large number of ultrastructural illustrations of some of the prominent pioneer authors included in the book, will provide ample information to the post graduate students and research scholars of insect neuroendocrinology as well as to those engaged in carrying out researches on comparative endocrinology, insect ethology, cell biology and developmental biology in various universities and acquaint them well with almost all the components of the neuroendocrine system in the active, inactive and various developmental stages, alongwith ultra-structural changes occurring in the cells and their organelles, that are associated with specific functions. I do hope that the book will stimulate the readers and serve as a ``text book cum basic research monograph'' on the subject. The book also contains a list of selected bibliography for a detailed study.

ISBN: 9788172333904 | 2005 | 444 pages

Agricultural Insect Pests and their Control

2nd Revised & Enlarged Ed.

V.B. Awasthi

This book has been mainly written keeping in view the requirements of the students of Agricultural Entomology of various Indian Universities. The book provides a brief account on the structure, function, development and metamorphosis, and a comprehensive account on almost all the methods of pest control. The nature of damage caused by the important insect pests of crops, their life histories and control have been described. It is hoped that the students of Agricultural Entomology, will find the book extremely useful. The book is also useful to the Plant Protection Departments and similar Institutions. The book will also be useful to those who are preparing for competitive examinations conducted by various central and state Government agencies for recruitment. Every attempt has been made to provide necessary information from the point of view of students of Agricultural Entomology, which is not available in other books.

Contents

ISBN: 9789386237125 | 2016 | 304 pages